Meliponiculture & Beyond in The Philippines

Author: Abu Hassan Jalil
Editor: David W. Roubik

IBRA
INTERNATIONAL BEE RESEARCH ASSOCIATION

Jointly published by: The International Bee Research Association,
a Company Limited by Guarantee, 1, Agincourt Street, Monmouth, NP25 3DZ (UK) &
Northern Bee Books, Scout Bottom Farm, Mytholmroyd, Hebden Bridge HX7 SJS (UK).

Obtainable from:
www.ibra.org.uk & www.northernbeebooks.co.uk

ISBN: 978-1-913811-14-3

Artwork by DM Design & Print
IBRA Proof Editor - Stuart A. Roberts

TITLE: MELIPONICULTURE AND BEYOND IN THE PHILIPPINES

Synopsis: This compendium is set in six parts covering the main Philippine regions. The first part is CALABARZON **Region IVA** which covers Cavite, Laguna, Batangas, Rizal & Quezon provinces. The second part is MIMAROPA **Region IVB**, and this covers Mindoro, Marinduque, Romblon and Palawan. The third part is Bicol **Region V** and the fourth is **Visayas Region VI, VII & VIII**. This part covers the islands of Panay, Guimaras, Negros, Cebu, Bohol, Leyte and Samar. The next part covers Davao **Region XI** in Mindanao. The vignettes and notes on each part are accounts of meliponaries and observations by the author from visits to more than 70 meliponaries throughout the Philippines over six years.

In the final part, 'Meliponiculture Distribution' in Central Luzon Regions I – III, and also the other regions not covered by the text, are shown in charts in the Appendices. The Addenda contain lists of beekeepers and meliponiculturists in the regions. The last section has excerpts from an advanced meliponiculture course.

All the parts together contain 312 sets of more than 500 images and about two dozen charts, maps and tabulations. it is hoped this can be a useful learning aid for the conscientious beekeeper intending to advance their knowledge and meliponiculture skills in and around the Philippine region.

Photographs And Illustration Caption Labels

Chart	Data and information charts.
Image	Photographs by others submitted in contributory articles and addenda
Figure	Photographs and illustrations by the author
Map	Geographical data or route indication
Table	All forms of tabulation

Abbreviations

AAA	Asian Apiculture Association
AHJ	Abu Hassan Jalil
ATI	Agricultural Training Institute
BEES	Bees for Environment & Economic Sustainability
BEENET	Beekeepers' Network
BBsU	Balay Bahay sa Uma
CAFGU	Citizen Armed Force Geographical Unit
CAS	Community Adoption and Strengthening
CBSUA	Central Bicol State University of Agriculture
CvSU	Cavite State University
DWR	David Ward Roubik
FPATF	Faith Paradise Agri-Tourism Farm
FEMACO	Federation of Marinduque Agriculture Cooperatives
ISAT-U	Iloilo Science and Technology University
LNCA	Leon Campus Alumni
MSEUF	Manuel S, Enverga University Foundation
NAP	Nora Alanza Peña
NMAP	Norwegian Mission Alliance Philippines
OMBKA	Occidental Mindoro Beekeepers Association
OMSC	Occidental Mindoro State College
OMNC	Occidental Mindoro National College
PHI	Philippines Honeybees Industries
SMAAC	Sierra Madre Agroecologists Agriculture Cooperative
STIARC	South Tagalog Integrated Agriculture Research Centre
UPLB	University of Philippines Los Baños

CONTENTS

PART 1 - CALABARZON

Luzon Photos: Fernando C. Belmonte Jr.

Preface to Part 1

This part is a meliponiculture tour of South Luzon and it was quite timely because although it was still in the wet season, we did not encounter much rain throughout. It took a couple of months of planning and communicating with beekeepers in Cavite, Rizal, Quezon, Laguna and Batangas Provinces. With the schedule set for a week of intense travelling and with the right company to travel with, the journey begins. The trip was successful with each stop occupant readily ferrying us from one stop to another. The whole journey saw a few different techniques of hive preparation and some peculiar vegetation provided for the bees. All in all, there were 13 stops in five provinces with a chapter for each stop plus two appendices.

The tour visited four cities and four municipalities. The cities are Lipa, Lucena, Tagaytay, and Tayabas; and the municipalities are Indang, Tablas, Alaminos, Mataas Na Kahoy, and Montalban if to include the starting point of the journey which will bring the total visited municipalities to five. Among the cities and municipalities visited, Lipa and Mataas na Kahoy are in Batangas province, Alaminos in Laguna province, Tayabas and Lucena are in Quezon province, Tagaytay and Indang are in Cavite province and Montalban in Rizal province. The trip covers six provinces in all. Of the six provinces, five consist the Region IV-A or CALABARZON which stands for Cavite, Laguna, Batangas, Rizal, and Quezon. Before its creation as a region, Calabarzon, together with Mimaropa and Aurora of Central Luzon, formed the historical region known as Southern Tagalog, until they were separated in 2002 by Executive Order No. 103.

The Philippines has two seasons, dry season and wet season. The dry season begins in mid-November and ends in mid-May, while the wet season begins in mid-May until mid-November, thus it gives six months period for each season. The dry season has two sub-types: the cool season from mid-November to February, and the hot season or summer which starts in March to mid-May. The cool season brings no snow except frequent frost on the highlands of the Cordillera region. It gives the country a milder and ideal climate than the rest of the year brought by a cold breeze of *Amihan* or northeast monsoon. However, the meliponiculture tour took place during October which is in the wet season but experienced less rain since the southwest monsoon or habagat that brings heavy rain from the West Philippine Sea has just ended.

Introduction to Part 1

Intending to explore "Island Meliponiculture", we set out to explore the largest main island of Luzon. Having covered Provinces like Laguna, Bicol and Sorsogon previously, we now decide to travel around South Tagalog. The vernacular names of the bees vary greatly from, region to region. In some instances, just monggo monggo is given to mean *Tetragonula* spp. In more common situations 'Kiwot' is given to *Tetragonula biroi* especially those with spiral brood arrangement and in other cases, the term 'Lukut' is used. Sometimes it is lukutan and in some villages, it's kilulut or kalulot and in some cases lib-og. We cannot attribute the vernacular names to a specific type or genus because they vary from island to island as the dialects of the Filipino language vary.

Southern Luzon Tour

JR Belmonte and Christ Mark Bergonio and Mrs Bergonio were there to greet me and usher me to Mark's farm in Indang Cavite. It was a long drive, which normally would be nothing to me but seeing that I just touched down with hardly any sleep, it was a bit overwhelming. Along the way, Christ kind of sensed my predicament and stopped at a convenience store where we had some coffee. Of course, all the while talking about bees.
We soon arrived in Cavite and headed for Los Pepe Farm. It was established by his wife's grandfather Pepe. It was handed down via his father-in-law.

Acknowledgements for Part 1

AHJ is grateful to Prof Dr Cleofas Cervancia and the organizing committee of the AAA2020 Philippines, 2nd International Meliponine Conference and the BEES2020 exhibition events. We gratefully acknowledge Peter Cervancia for including us in the display of his products at the BEES2020 event.

AHJ thanks Prof Dr Deborah Ruth Smith for her contribution of excerpts and images from her "Genetic Diversity of Stingless Bees" paper. Sincere appreciation to Ms Luz Z Gamba for the hospitality extended during the Balay Buhay sa Uma visit in Bulusan, Sorsogon.

The authors thank Lee Gaitana for hosting the Advanced Meliponiculture Course at Pia's Bee Farm and seeing the needs of the course.

Distribution of Meliponiculture activities in CALABARZON

Meliponiculture Distribution in Central Luzon Region I - III

Chart 1 Distribution of Meliponiculture activities in Northern & Central Luzon Region I - III

List of Meliponiculture activities in Northern & Central Luzon Region I - III

1. Leo Esnara Kimbungan, BSU, Benguet
2. Baltazar Kiat-ong, La Trinidad, Benguet
3. Arthur Benosa, Sagunto, Sison, Pangasinan
4. Texter Taclobao, Quezon city
5. Elmer Portugal, San Juan, Botolan
6. Rommel Mngalindan, Quezon city
7. JR Belmonte, Polilo - Montalban
8. JR Belmonte, Bataan
9. Mark Salinas Esmana, Mariveles, Bataan
10. Clent Tyron, Quezon City

- All data gathered in this Appendix was up to the end of 2019.

Meliponiculture Distribution in CALABARZON Region IVA

Chart 2 Meliponiculture activities in CALABARZON Region IVA

List of Meliponiculture activities in CALABARZON[1] Region IVA

1. Ilog Maria, Silang, Cavite
2. Hardin Sa Parang, San Mateo, Rizal
3. JR Belmonte, Montalban, Rizal
4. Bernard Tuzon Anciado, San Benito, Alaminos, Laguna, Luzon
5. Melvin Monzon, Luisiana, Laguna
6. Christ Mark Bergonio, Los Pepes, Indang
7. UPLB Transit Depot, Pasay City
8. Kota Paradiso, Mabitac
9. Sayo Nora Farm, Majayjay, Laguna
10. Sol Benedict, Mt Banahaw, Delores, Quezon
11. CvSU Meliponary
12. Marvin Creencia Manila, Los Baños
13. Christmas Osuna, Nagcarlan, Laguna
14. Girsky Villasante Anda, Alimonan, Quezon
15. PJ Plantation, Anos
16. UPLB Campus Grounds, Los Baños, Luzon
17. Lee Gaitana, Batangas
18. Haypee Piñero Maliwa, Laguna
19. Sandoval Farm, Gibanga
20. Victor Rea Oribe, Lucena
21. Niño Jhammel Loba, Gumaca, Quezon
22. Rehab Transit Bee Prog UPLB
23. Percival Alvarez, Bay Laguna
24. Cudia Farm, Mataas na Kahoy, Batangas

[1] CALABARZON Region IVA
Among the cities visited, Lipa is a city in Batangas province, Alaminos is in Laguna province, Tayabas City is in Quezon province, Tagaytay City is in Cavite province, Mataas na Kahoy is in Batangas province so with Romblon the trip covers 6 provinces. The prelude would have covered Rizal it starts from Montalban, Rizal, and the feral hive transfer article is in the appendix. so all in all 5 provinces of CALABARZON covered plus Romblon. Mataas na Kahoy, and Alaminos are both municipalities (not a city). All are in the region IV-A (CALABARZON) CALABARZON means CAvite, LAguna, BAtangas, Rizal, QueZON. Prior to its creation as a region, Calabarzon, together with Mimaropa and Aurora of Central Luzon, formed the historical region known as Southern Tagalog, until they were separated in 2002 by virtue of Executive Order No. 103

Chapter 1 – Cavite Province

1.1 Christ Mark Bergonio, Los Pepes Farm – Indang, Cavite

The farm is situated in an exclusive remote plot that appears to have natural vegetation and without other agricultural neighbours that may be using pesticides. Los Pepes Farm comes from Kathleen's grandfather named Pepe (Jose). It was established the year 2012. He continued from Roberto P. Alano his father-in-law.

The farm has a decent collection of feral colonies gathered from the nearby area. Feral colonies are especially of concern as they represent the endemic species of that region.

One could find almost every kind of housing applicable. He probably tried many methods over the years as can be seen from the clay pot trials, coconut shell trials and boxes of many dimensions.

Figure 1 There were not many plants and vegetation, however, almost every type has at least one representative.

Figure 2 With Chris Mark at Los Pepes Meliponary

There were not many plants or vegetation, however, almost every type has at least one representative. There are guavas. pomegranate, soursop, jackfruit, banana and some assorted shrubs and groundcovers. An

Figure 3 It appears Mark has inherited almost every conceivable type of bee housing as we see an assortment of containers.

umbrella is used in its literal sense as 'protection from rain'. This farm is an intriguing place to visit for stingless bee lovers. The hives are established, and many years of character abound in the nest entrances. The 'Beescape' has much room for development if new hives are to be added.

Novelty box hives of old-school beekeeping techniques. We see a baggage trunk, an armchair, a truck, a boat and possibly a community hall. The elaborate and ornate nest entrance depicts box construction from a decade ago. The resilient imagination and creativity of the original owner have to be from an older generation.

Figure 4 Novelty box hives of old-school beekeeping techniques

Figure 5 The younger generation has explored much in innovative products and derivative products.

The younger generation has explored much in innovative products and derivative products. We were only allowed to depart with a whole load of goodies in our hands.

1.2 Michel Bono, CvSU

While in Cavite it was suggested I visit the Cavite State University to meet up with the University President. Mark subsequently brought me to meet Madam Michele Bono, head of the CvSU Bee program. Maam Chel (as casually called) introduced me to Dr. Robles where we exchanged pleasantries and I presented one of my books and a flyer for the upcoming 2nd International Meleponine conference next February.

Figure 6 Cavite State University CvSU Meliponary (inset: Presenting a book to Dr. Robles President of the University)

I was then brought to visit the Meliponary where I noticed it was in a mini coconut plantation and dragon fruit trees abound. The boxes were a bit haphazardly constructed and finished. I advised giving it a coat of Turpentine Extracted Propolis (TPE) to improve the box hive's durability and lifespan. I further imparted some tips on Meliponary Management and we discussed insect ecology with some of the lecturers.

Figure 8 A visit to the Meliponary and a discussion with the Bee Program Staff members.

Figure 7 Remote sensors innovated by students of the engineering faculty

It was good to know that the engineering faculty had set up some temperature monitoring sensors. However, it was not in working condition when I visited.

See Appendix 2 for the full report on CvSU Bee Program initiatives.

Figure 9 With some final words we parted after lunch, together with Bernard Anciado

1.2.1 Addendum - Community Adoption and Strengthening of the CvSU Bee Program (CAS Bee)

Republic of the Philippines
CAVITE STATE UNIVERSITY
Don Severino De Las Alas Campus
Indang, Cavite

Cavite State University Bee Program

1.2.1 Establishment of meliponary

CvSU BS Biology Batch 2001 alumni started their meliponary at Brgy. Tambong Kulit, Indang, Cavite with 10 starter stingless bee colonies. Brgy. Councilor Aniceto Herrera of Calumpang Lejos also started his meliponary with 8 colonies for crop pollination.

Image 1 Establishment of meliponary

1.2.2 Paper Presentations, Awards and Recognition

The Beekeepers' Network Philippines Foundation Inc. (BEENET) gave a certificate of recognition to the CvSU Bee Program during the 25th

Image 3 Donation of 20 hive boxes for the "Iwi System/ Bahay para sa Lukut" Project from CvSU BS Biology Batch 2001 Alumni

Image 3 Mr. and Mrs. Bergonio of Los Pepes Bee Farm showcasing their stingless bee products during an exhibit in Indang, Cavite

BEENET Philippines Conference last July 24-26, 2019, for extending beekeeping activities in the province of Cavite and for the program's support in the Philippine beekeeping industry. Also, the CvSU Bee Program's stingless bee honey won first place in the "Best Tasting Honey" contest during the said conference, which was attended by beekeepers in the Philippines. In 2019, 2017, and 2015, the CvSU Bee Program won the Best Extension Paper, Best Poster and 3rd Place Best Paper, respectively, in the CvSU Annual Research and Extension In-House Review. On October 10-11, 2019, this program also won Best Extension Paper in the 32nd Regional Symposium on Research, Development and Extension Highlights of the STAARRDEC.

1.2.21 Donations for the Iwi System/ Bahay para sa Lukut Project (Component 2 of the CAS Bee program)

CvSU BS Biology Batch 2001 alumni donated 20 hive boxes to the CvSU Bee Program last Sept. 3, 2019, which would be entrusted to the trainee adopters of Brgy. Calumpang Lejos, Indang, Cavite and Muzon Farmers' Association. This would serve as their start-up hive boxes to establish their bee farm/ meliponary.

1.2.2.1 Support to entrepreneurs - Kapit BEEsig Project (Component 3 of the CAS Bee Program)

The CvSU Bee Program is continually supporting its trainee-adopters who ventured into business. Mr. Christ Mark Bergonio and Mrs. Kathleen Bergonio of Los Pepes Bee Farm in Calumpang Cerca, Indang, Cavite, engage in bee product business like selling propolis spray, pollen, propolis, honey and stingless bee colonies. They sell their bee products nationwide. In support, the CvSU Bee Program promotes its products and sells them during exhibits and trade fairs. Stingless bee colonies are also being sold by Los Pepe's farm to interested beekeepers.

CvSU Bee Program also provides linkages and networks for the beekeeping and business' needs of the clients.

1.2.3 Beekeeping in the Academic Community

Classes on Biodiversity, Biotechnology, Ecology, and Graduate Seminar in undergraduate and graduate programs were given demo lectures on the importance of bees and beekeeping technologies. Multi-disciplinary bee-related research was conducted by faculty and students, which was also disseminated in the community during the training seminar. Beekeeping became one of the electives in the BS Biology Program curriculum, titled BELE – Introductory Beekeeping Course.

Image 4 Demo-lecture on the importance of bees and introduction to beekeeping

1.3 Ilog Maria Honeybee Farm

The next day we were brought on an excursion to Silang, Cavite to visit Ilog Maria Honeybee Farm. https://www.ilogmaria.com/

Ilog Maria Honeybee Farm km. 47 Aguinaldo Highway, Lalaan 1, Silang, Cavite, Philippines 4118

1.3.1 Etymology: Ilog Maria, Ilog means River and I was told by the locals that the river's name was dedicated to the Virgin Mary, hence Ilog Maria.

1.3.2 Getting there: Similar to previously, the mode of transport was vans and this time it was only 2 vans as it included only the delegation of the conference speakers. It took one and a half hours to get there.

It is a large spread for a Bee Farm, and we were utterly impressed. They also produced many value-added bee products and their retail store provides items for local purchasers as well as tourists.

Figure 10 Ilog Maria signage

Figure 12 Extension courtyard underway

Figure 11 The Bucari Pine left and Bagras the Rainbow Eucalyptus

Figure 13 Hives well-spaced in the Back Meliponary

Chapter 2 – Laguna Province

2.1 San Benito – Bernard's Transit Home, Alaminos Laguna

The destination was Bernard's home and farm in San Benito, Alaminos where we arrived amid an hour of some Filipino humour from JR Belmonte and Bernard. I was kind of a blur initially, but I guessed I caught on.

Figure 14 Residenza de Alaminos Brgy 2 Población Alaminos, Laguna, Bernard's home away from home. In front of Bernard's '2nd home,' we were greeted by a unique replica of a bee on dummy giant sunflowers. Besides this was a pot of Nepenthes sp. (Monkey cup plant) which was occupied by a Tetragonula biroi colony.

In front of Bernard's '2nd home,' we were greeted by a unique replica of a bee on dummy giant sunflowers. Besides this was a pot of *Nepenthes* sp. (Monkey cup plant) which was occupied by a *Tetragonula biroi* colony. This, he said, was the result of a split he did some time back. The original split colonies are on the windowsill, while this one occupying the *Nepenthes* plant was a fluke of a natural occurrence.

The original split colonies are on the windowsill, while this one occupying the Nepenthes plant was a fluke of a natural occurrence sa Residenza de Alaminos Brgy 2 Población Alaminos, Laguna and the Location of the Farm is Brgy San Benito Alaminos, Laguna...his other colonies are scattered in the town of Alaminos.

This one I believe to be a *Tetragonula* cf. *sapiens* nest having interspersed cells with connectors he calls the monggo monggo.

Figure 15 Tetragonula cf sapiens nest having interspersed eggs with connectors.

From there we travelled to a remote plot where his actual farm is. We had to thread a distance through some farmland. There was a coconut plantation and a fruit orchard. Through that, we came by some box hives arranged in groups. Fortunately, he had the foresight to space the groups out and not cluster them together.

We also met with a friend of his Mr. Morota who was quite happy to meet us. Mr. Lui Morota has dabbled in some propolis product development and so was quite eager to share his ideas.

His actual bee farm is a way farther in, which is quite wise as the neighbouring farm is quite active.

As we arrived at the bee farm proper, we were met with another group of beekeeper friends i.e. Haypee Piñero and company namely, Annie Joy de Roxas and Jesus Lanuzga. It was good meeting more people with the same interests.

Figure 16 Bernard Anciado in some of his satellite farms

Figure 17 His actual bee farm is way farther in, which is quite wise as the neighbouring farm maintenance activities may upset the bee's foraging equilibrium.

Bernard Tuzon Anciado Bee farm

It was refreshing to see the bees spaced out in a vast area and foraging freely with minimal competition. Bernard has been keeping stingless bees for a few years, spending only his off-duty hours on his bee farm. He has a quite regimented approach in his beekeeping (rather obvious in a uniformed officer's demeanour). His older mother's hives are clustered on racks while his split colonies are positioned on singular stands and almost equally spaced.

Some of his boxes are painted in whitewash and he showed us a newly split hive, a well-established hive and an ultra-productive one in which he recently harvested up to 3 kilos of honey in one season. The harvesting of honey is one season per year i.e., between the *Amihan* and *Habagat* (wet and dry respectively)[2] seasons.

We had a very fruitful discussion about increasing honey yields and possible steps in overcoming setbacks. After a group photo session, we left for the next destination.

Figure 18 Discussions regarding increasing honey yields and possible steps in overcoming setbacks

[2] The Amihan wind (from the North) lasts from October to the end of March, while the Habagat (from the South) rules the other half of the year.

2.2 UPLB, Los Baños, Laguna, Luzon

Alex Fajardo showed me a large rain tree on the Campus grounds of UPLB, Los Baños, Laguna, Luzon

Figure 20 Some of the elaborate nest entrances on the rain tree.

Figure 19 The rain tree (Samanea saman) at The University of Philippines Los Baños (UPLB) has 27 colonies of Tetragonula biroi

The rain tree (*Samanea saman*) at the University of Philippines Los Baños (UPLB) has 27 colonies of *T. biroi* with each colony being 1.5m apart from the other.

Several rain trees found on the campus grounds have lukut colonies and oddly being approximately 74 m apart from each tree attested by Alejandro Fajardo Jr.

7

yr.

Figure 21 Pollen specimen charts by Manila-Fajardo & Cervancia 2003 with 30+yrs collection in UPLB.

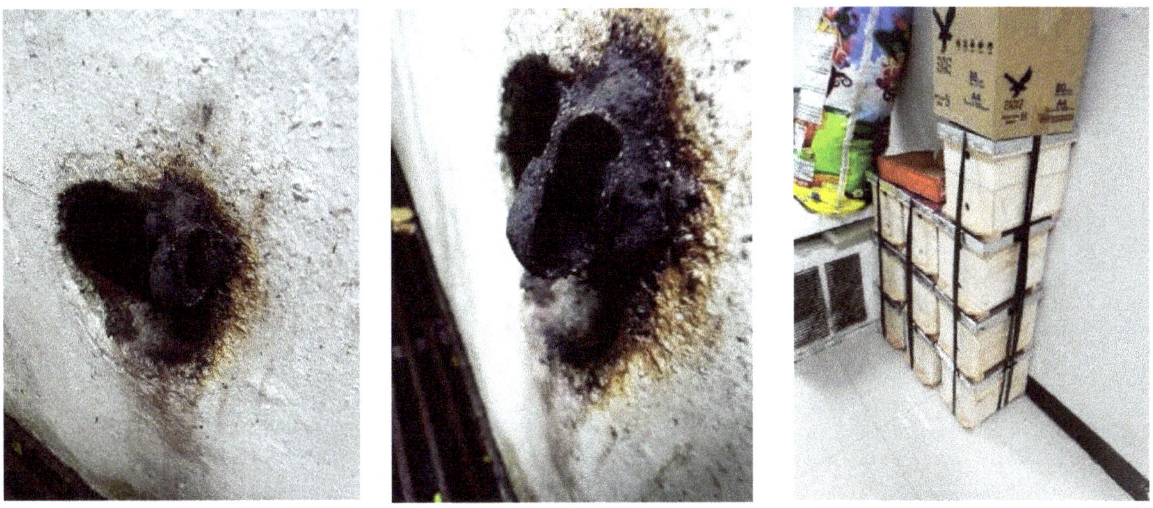

Figure 22 Nests of Tetragonula cf. sapiens found in the brick walls of the Lab building possibly that of Tetragonula fuscobalteata. Box Hives in stock ready for use with hives in the UPLB Bee yard.

Colonies in the UPLB bee yard starts with wild colonies inducted into coconut shell hives. (The coconut shells are placed in layers on top of each other in alternating orientation.) Once established with the brood chamber in the nucleus, there are honey pots in the preceding layer (either the bottom or side layer) after which comes the layer with pollen pots.

Figure 23 Indications of a weak colony are seen in reduced congregations outside the hive and brittle, greying dry nest entrances flight tube extensions

During the transfer process, the coconut shells are peeled off and depending on the season, the pollen and honey pots may or may not be used. Usually, if it is the rainy season, some of the pollen pots may be used, because there will be very little pollen that bees can gather in a rainy season. It is not recommended to add pollen if the colony is weak, nor is it recommended to do transfers or splits when they are weak. Chances are they may not survive due to overwhelming parasites and pests that come into the hive.

Indications of a weak colony are seen in reduced congregations outside the hive and brittle, greying dry nest entrances flight tube extensions, due to not being maintained or reduced maintenance. Another factor is the aperture of the flight entrance tube. If the entrance aperture is small or only fits one bee, then it is a sign of the colony deteriorating. Strong colonies will have pliable entrances and a heavy congregation outside the hive and have large apertures.

In the bee yard, they have hives for training as well. Sometimes intense battling occurs among hives especially if a new hive is brought into the yard. If colonies are relatively the same size or strength, they will shy away from fighting to avoid disastrous casualties. Colonies are brought in any time of the year regardless of season, but the tendency is to bring in hives of the same size and strength and not bring in weak hives. Weak colonies will tend to die out and battling swarms[3], regardless of the season, happen daily within the yard of 24 colonies. However, the fighting swarms occur from 10 am to noon and again in the early evening around 3 p.m., Alejandro Fajardo Jr. explains that the swarms are not for reproduction but involve competition (or possible mating aggregations of mostly males).

[3] Editor's note: everything indicates the beekeepers put too many hives close together in a small space, and planting a few rows of flowers nearby basically does nothing but make the beekeepers feel better.

Figure 25 Initially in coconut shell hives then transferred into boxes.

Fortunately, the yard does not get wild monkeys or wild boars but gets the occasional group of naughty vandalizing young boys.

Initially in coconut shell hives then transferred into boxes. UPLB Bee Program Yard and the UPLB Bee Program Leader then was Dr. Amee Dupo.

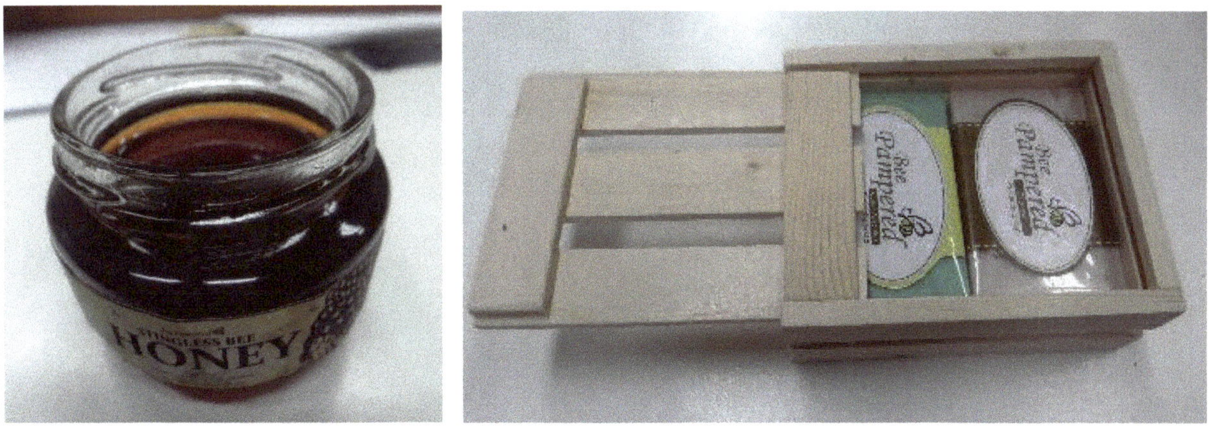

Figure 26 Left: Bottled Dark wild Honey; Right: Propolis soap products in wooden packaging

Display of Honey Cider Wild Lukut honey & *Apis mellifera* Honey, Honey Mead - Wooden Pack

Figure 27 Left: Honey Cider; Middle: Wild Lukut honey & Apis mellifera Honey; Right: Honey Mead ib Wooden Pack

2.2.2 Transit and Rehab Depot of the Bee Program for UPLB

On the way back to the Manila Airport, together with Prof. Cervancia, we visited this facility.

Figure 28 Transit and Rehab Depot of the Bee Program for UPLB, Pasay City, Manila

The rehab centre houses a wood workshop for building hives boxes and feeding stations for bee recuperation. Feed is diluted spilt honey in drips from inverted bottles over flannel sheets.

Figure 29 Feeding stations for bee recuperation. Diluted honey dripped from inverted bottles over flannel sheets.

Over a hundred fern hives salvaged after the recent typhoon are brought here in transit and rehabilitation, while a wood workshop is dedicated to building hive boxes for the ultimate transfer of the fern colonies into the box hives.

Figure 31 During the recent typhoon, hundreds of fern hives were salvaged from the wild surroundings and brought to this transit & rehab depot for recuperation and ultimate transfer to box hives.

Figure 30 The rehab centre houses a wood workshop for building hive boxes

2.3 Sayo Nora Farm in Majayjay City, Laguna

From Los Baños, we travelled to Majayjay City, Laguna Visited Sayo Nora Bee Farm in Majayjay on 21st Sept. 2017. Run by Patricia Bravante and Family.

Figure 32 Sayo Nora Bee farm Initial spread of 1,500 hives.

Figure 33 Four hives on a balanced metal stand anchored in the ground.

Losses of 45% occurred initially due to hive congestion and poor placement. Now 4 hives are positioned on balanced metal stands anchored in the ground.

Figure 34 Sayo Nora's products. and the Sayo Nora farm management with consultants

Sayo Nora farm management with consultants from left, Ann Marie Bravante (daughter) Mr. Bravante (father), Alejandro Fajardo, Abu Hassan Jalil, Marvin Creencia Manila and Patricia Bravante (Daughter and Manager). Displayed here is a part of Sayo Nora Bee farm's products.

Figure 35 Tetragonula biroi box hive with spiral layered brood cells

2.4 Three Miel Bee Farm

I visited Three Ml Beefarm in Los Baños and I arrived by nightfall due to the whole day of travelling around. My experience there was limited to what we could view with a torch light and spill-over lights from the house. It is owned and run by Alejandro C. Fajardo Jr. and his wife Dr. Dahlen Manila-Fajardo.

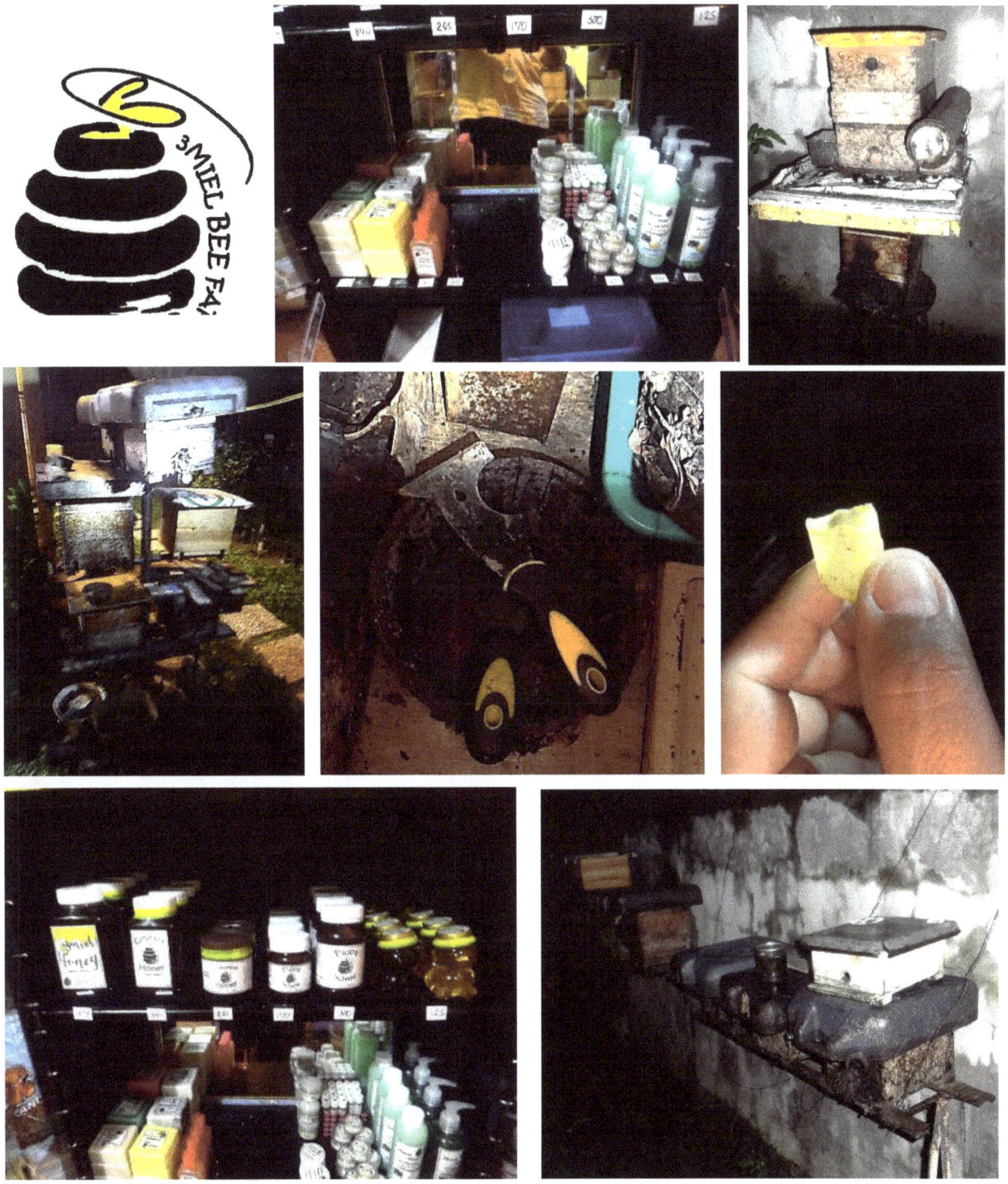

Figure 36 3 Miel Bee farm in Los Baños, for example, prying tools and sample wax moulded storage pot for the bees.

Figure 37 60 odd colonies with much experimentation in Tetragonula Meliponiculture.

Three Meil Beefarm houses about 60 odd colonies with much experimentation in *Tetragonula biroi* Meliponiculture. I captured a few microscopic images with my portable digital microscope. Alex told me that the species *T. biroi* was verified by the Filipino taxonomist Balthazar.

2.4.1 Addendum – 3Miel Bee Farm Write Up

By Dr. Dahlen Manila-Fajardo

3Miel Bee Farm is a young beekeeping company, legally established in 2016; but is handled by people with more than 25 years of beekeeping experience, including commercial beekeeping in Australia and New Zealand. We have been engaged in bee research and extension for many years with the UP Los Baños Bee Program. We have been involved in training a lot of people from different walks of life on bees and beekeeping from all over the Philippines. With the program, we were also able to write a lot of journal articles; manuals, guides and semi-technical works; and developed beekeeping technologies. After serving with the university, we finally decided to give the business side a shot. The knowledge and skills we learned from the program gave us an edge in doing training and innovations.

Our name, 3miel, stands for the honey harvested from three species of bees: *Apis mellifera*, *Apis breviligula* and *Tetragonula*, which we always have. We also maintain and handle the three species of bees on our small farm. Incidentally, it also pertains to the three kinds of honey in our lives – our lovely kids.

Our company is composed of a 3-man team who consider themselves more as bee scientists rather than just beekeepers alone. This keeps our management and other activities dynamic as we incorporate basic technical know-how in our applied beekeeping. Alejandro "Alex" C. Fajardo Jr. is our entomologist specializing in bee pathology and taxonomy. He is the designer of the well-known UPLB TPH-1, the stingless beehive design used in most Meliponaries in the country and utilized as the base design of new hive box innovations. His spouse, Dr. Analinda "Dahlen" C. Manila-Fajardo, is a pollination ecologist and an active advocate of enhancing honey quality, especially those coming from stingless bees, having done honey and pollen analyses for many years. Marvin C. Manila, a technical educator and fabricator, is doing most of our beekeeping activities. He is also doing instrumental insemination for the company's queen production.

We offer technical services such as consultancies on pest and disease management, pollination services, colony management, basic information about bees and beekeeping, pollen and bee identification, basic honey and palynological analyses and student research advising. We are the only private beekeeping outfit that performs bee removal-relocation services. We remove bees and return them to the mountains. We are the only private beekeeping outfit allowed to release Philippine Giant Honey Bees at the Mt. Makiling Forest Reserve.

We also offer beekeeping skill development from stingless bees to honeybees, bee product processing and development, management and economics, carpentry, and pollination with bees. We also produce several basic and value-added products from bees. The basic products we sell are honey, pollen, beeswax and propolis. Some of our value-added products where we incorporate the basic products are shampoo, propolis ointment, massage oils, lip balm and soaps.

We continue to innovate to cater to emerging challenges in beekeeping; most notably in hive design development and hive management for stingless bees and *Apis cerana*, in collaboration with our beekeeping partners. We forge partnerships with the private sector and select government agencies in promoting the commercialization of bee products and uplifting the beekeeping sector from a backyard activity to a full-fledge agricultural industry.

Currently, our place is 63km away from Manila. We are only a km. before the entry of UP Los Baños. Our office and depot are at Mt. Kitanglad St., Umali Subdivision, College, Laguna. Our colonies are spread out, as we practice migratory beekeeping in several places in Laguna such as Liliw, Bay and Calauan and Tiaong, Quezon.

We hope that we establish proper protocols for honey harvesting, processing and pasteurization, and storage, especially for various stingless bee species in the Philippines. And in partnership with pertinent government agencies, we hope to develop national standards for stingless bee products, colony health and GMP. It is our vision to create an Api/Meliponitourism site to feature Philippine bees and products from all over the country. We will make it a site where members of BEENET Philippines and its allied associations can display and sell their products, bearing their labels.

2.5 AAA 2020 Technical Workshop at Mabitac

The "International Meliponine Conference and Asian Apicultural Association (AAA) Philippines Symposium on Pollinator Conservation" officially came to an end on 28 February 2020 at the Kota Paradiso in Mabitac, Laguna.

2.5.1 Etymology: Mabitac was an excellent hunting ground for wild game three centuries ago. Native hunters used numerous cave-ins or

Figure 38 Having visited Kota Paradiso in Mabitac, we made our way back to UPLB

trap-ins called "bitag" in the local dialect. Hence, the place was referred to as "Mabitag" meaning "a place with many traps".

2.5.2 Getting There: The whole conference delegation was transported by a convoy of vans from UPLB It was more or less a 2-hour trip. We started at about 6:30 a.m. and arrived at Kota Paradiso, Mabitac at about 8:30 a.m.

2.5.3 **Last Day of the AAA 2020 conference** In collaboration with the Department of Agriculture-Bureau of Agricultural Research (DA-BAR) and organized by the AAA through the University of the Philippines Los Baños (UPLB) Bee Program– the last day of the conference cum symposium featured a technical tour, roundtable discussion, and closing ceremony that was attended by members of the national and international apiculture community.

Figure 39 A technical workshop for the conference delegates at the stingless bee farm of Kota Paradiso, Mabitac, Laguna

Chapter 3 – Rizal Province

3.1 Pictorial Report* by JR Belmonte

Five months feral colonies were relocated in a sack

This is the entire hive inside the sack. it's from the dried leaves of ferns (pakpak lawin/dapo).

In the following pictures, you will see the cerumen bits that the bees gather from the other hive that I transferred previously. they stored it near the entrance tube.

Figure 40 This is the entire hive inside the sack. it's from dried leaves of ferns (pakpak lawin/dapo).

Also, near the entrance tube, they store fresh resins. Here you can see it is greenish; the cerumen bits from the above pictures are orange-brown.

Figure 41 The entrance tube stores the fresh resins. Here you can see it is greenish. The cerumen bits from the above pictures are orange-brown.

I tried to remove first the dried leaves and debris that are attached to the hive. I cleaned the bottom part first and I found a termite nest attached to it.

The termite colony co-existed with the *T. biroi* colony in a single hive.

I finally found the brood and I exposed a little of it so I would know how it is positioned inside the hive. We can see in the above pictures that the termite colony is just approximately 2 inches away from the nearest brood comb (or plate) of the *T. biroi*. the only boundary is a dried leaf of the fern and a layer of honey underneath the bottom part of the brood plate.

Figure 42 I tried to remove first the dried leaves and debris that are attached to the hive.

Figure 43 The termite colony co-exists with the T. biroi colony in a single hive.

After finding the brood, I started to clean the hive by removing first the hard propolis and other debris near the entrance tube.

I removed the outer dried leaves of the fern just immediately below the entrance tube and I saw, in the above picture, the concentration of pollen pots with few honey pots.

Figure 44 At the side near the brood plate I saw this group of honey pots

At the side near the brood plate, I saw this group of honey pots.

And... where I'm pointing with my finger is the place where the group of honey pots were located. on top is still the pollen pots near the entrance tube. below the pollen, pots are the slightly exposed brood (above picture).

Here I'm beginning to expose the brood. the concentration of pollen pots is still seen on top.

Figure 45 Here I'm beginning to expose the brood. The concentration of pollen pots is still seen on top. Here we can see the queen brood cell. The entire brood is now completely exposed.

Here we can see the queen brood cell. The entire brood is now completely exposed. Surrounding the brood is a mix of pollen pots and honey pots (above pics).

Next, you will see the natural division of lower brood combs and upper brood combs. I wasn't able to measure it but the lower brood has 12 combs while the upper group has 4 to 5 brood combs.

Figure 46 Here you can see the natural division of lower brood plates and upper brood plates. I wasn't able to measure it but the lower brood has 12 plates. While the upper group has 4 to 5 brood plates

Here I already have transferred the lower group of brood combs to the bottom hive box (treated recycled hive box), and the top group of brood combs to the upper box. (Above pics)

After transferring, this is an old hive box that I re-used. All brood chambers have an entrance hole on each one (above pic).

Here is the picture of honeypots underneath the brood combs that I removed.

3.2 Conclusion: pollen pots are the majority on top of the brood, near the entrance tube, and outer dried leaves. honeypots are concentrated underneath the brood and at the side.

Figure 47 Here I already have transferred the lower group of brood combs to the bottom hive box (treated recycled hive box), and the top group of brood combs to the upper box. After transferring, this is an old hive box that I re-used. All brood chambers with an entrance hole on each one. Here is the picture of honeypots underneath the brood combs that I removed.

Chapter 4 – Quezon Province

4.1 Pinky Sandoval - Gibanga, Quezon

Bernard Anciado brought us to bypass San Pablo city and somewhere after Candelaria we stopped to rendezvous with Dr. PJ de Castro. We soon met with Doc PJ who brought us to Gibanga to Maam Pinky Sandoval's Place.

After the sumptuous table spread and seeing that Pinky's place was a lovely farm spread, we ventured into the dark with torchlights to view her meliponary.

It was dark but somehow, we all saw the 'light'. We trod carefully and made our way to her meliponary. We did get to have an idea of the expense of the meliponary despite the darkness.

As it was getting late into nightfall, we decided to take leave and head for Dr. PJ de Castro's place.

48 After Dinner at Pinky's beautiful courtyard patio we visited her bee farm by nightlight.

4.2 PJ Farm and Plantation, Anos Tayabas

We arrived at Dr. PJ de Castro's house where we sat for a chat in his honey processing setup on the ground floor of his house. Later we put up for the night in the guest room on the 1st floor.

Figure 49 Dr Pj's office cum Honey Lab on the ground floor of his home, with his Honey extractor invention.

A very innovative contraption for the squeeze and strain method of honey harvesting, it uses a car jack to press upwards to a large piston that compresses without manual force. The squeeze is effective but the honey should not be in contact with metal parts for too long. The draining and cleaning should be done immediately as the acidic properties of honey may have adverse effects if left too long in the device.

The next morning, he took us to his plantation and farm in Tayabas. The farm was about 15 min. away and we got to see the village life along the way. Although it looked more like a cowboy's ranch, the horses were beasts of burden. They had metal baskets hung beside them for carrying coconuts.

Figure 51 The coconut 'cowboy' stables that we passed as we approached his farm and plantation.

Figure 50 The Hacienda is in progress and the feature will complete the plantation amenities.

4.2.1 PJ's Plantation and Bee farm – Anos, Tayabas

Figure 52 Dr. PJ's Plantation and Bee farm

Figure 53 Well-established coconut shell colonies.

Figure 54 Floats used as hive housing hanging from roof eaves

His plantation was like a ranch the way I see it. There was a 'Hacienda' with a manmade pond and a rest hut in the middle of the pond.

In the yard were clusters of coconut shell hives with colonies that have adapted well to the surroundings. Below are some photos of colonies that may have taken years to establish.

Hanging from the roof eaves are what looked like balls to us initially, then on closer inspection, we found that they were wharf or jetty floats (buoys from Polillo Island). Dr. PJ had cut a window flap at the back of the float and inserted the bee colony. What an ingenious way to house stingless bees and they seem to adapt well. The adaptation to the float was quite expansive as they made elaborate and intricate entrances. We then completed the tour of his farm and some sharing of meliponiculture techniques.

4.3 Victor Rea Oribe – MSEUF, Lucena City

We had the opportunity to visit with Dr. Victor Rea Oribe of Manuel S. Enverga University in Lucena City.

Figure 56 We had the opportunity to visit with Dr Victor Rea Oribe of Manuel S, Everga University in Lucena City.

Figure 55 Peculiar nest entrances of Tetragonula biroi. yellow in colour.

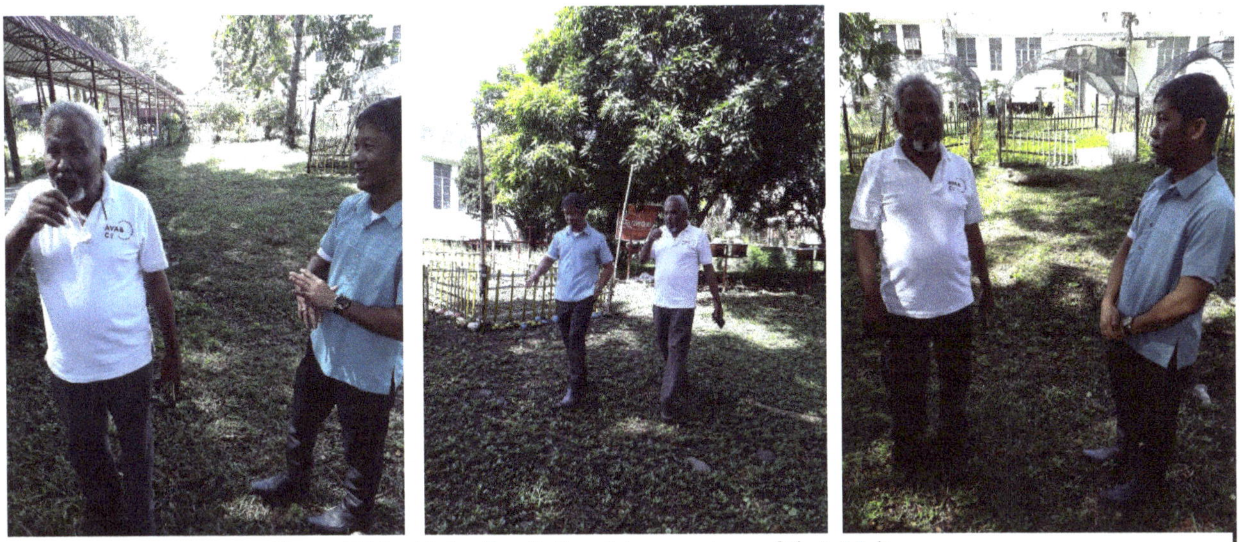

Figure 57 Dr. Victor Rea Oribe giving us a tour of their Meliponary.

The visit to the Bee Farm and Meliponary was quite interesting because we saw some very peculiar nest entrances of *Tetragonula biroi.* The entrances were yellow.

.I explained that from my experiences with varied colours of entrances, they were influenced by the resin of the tree where they were obtained. However, the internal structures and propolis remain a consistent dark brown colour. I suggested he document that phenomenon if similar.

Above are photos by Dr. Victor Rea Oribe of flowers visited by bees in his collection.

Figure 58 Fabaceae & Asteraceae with foraging bees photos courtesy of Dr. Victor Rea Oribe

We then headed off to the next destination, the South Tagalog Integrated Agriculture Research Centre (STIARC) in Lipa City, because we were told the coop members were already there waiting.

Figure 59 photos by Dr. Victor Rea Oribe of flowers visited by bees in his collection.

Chapter 5 – Batangas Province

5.0 South Tagalog Integrated Agriculture Research Centre STIARC

Lipa City, Batangas Province

5.0.1 Getting there: The drive from Lucena, Quezon Province to Lipa City took 2 hr. and finally we arrived at the South Tagalog Integrated Agriculture Research Centre STIARC

5.0.2 Etymology: Lipa city -. Batangueños from the early years had their settlement in Bombon Lake and began dispersing to other places when the volcano erupted. While a group of people was moving to another settlement area, the image of St. Sebastian was stolen from them and later on was found on a tree called "lipa". It is named after a species of a tree in the Urticaceae family, called lipa (*Dendrocnide meyeniana*), which is known for its twigs' stinging trichomes.

5.0.3 Etymology: Batangas City - Spanish plural form of the Tagalog word *batang*, meaning "log", about the trunks of logged trees that used to be floated down the Calumpang River which runs through the town (now city) of Batangas. Originally the name only referred to the town, but as with many other provinces created during the Spanish colonial era, the name of the capital town was applied to the whole province. Other former names of the Province that reflected the location of the administrative capital include Balayan, Bombon (a settlement on the shores of what is now Taal Lake, destroyed by a volcanic eruption and later re-established as the town of San Nicolas), and Comintan (after the settlement of Kumintang, now part of Batangas City).

5.0.4 **Meeting** with STIARC management representatives (Maria Riza Panopio - STIARC) and Batangas Beekeepers Co-operative (Cecilia Katigbak De Jesus @ Bee Merry, Manager of Lipa Beekeepers Coop & Owner of The Beehive Restaurant)

At the request of Bee Merry, we met with about 7 members of the cooperative and STIARC representatives. I showed the derivative stingless bee products that I brought. I briefed them on the potential and possibilities of developing many by-products and derivative products from Stingless beekeeping. I also showed the books, DVDs and charts that were brought. The meet was to discuss future courses and demonstrations on advanced Meliponiculture and derivative product programs.

We were offered sleeping quarters for the night and we took the opportunity to freshen up before moving on to Pia's Garden in Tambo for lunch and a farm visit the next day.

Figure 60 South Tagalog Integrated Agriculture Research Centre STIARC. we met with about 7 members of the cooperative and I showed the derivative stingless bee products that I brought together with the CDs and charts.

5.1 Lee Gaitana – Pia's Garden, Tambo, Lipa

We made our way to Tambo to meet Lee Gaitan who prepared lunch for us. Lee's bee farm is a garden setting that he calls Pia's Garden.

Figure 61 Pia's Garden with Lee Gaitana and Marvin Creencia Manila was also there to meet us.

His garden was truly like a page out of my "Beescape" book. I commented on the beauty of his garden and he admitted that he followed much of the principles laid out in my book. I was very proud and decided to spend a bit more time appreciating his work.

Pia's Bee Farm derived its name from the beekeeper's daughter. It is located on a plateau of 1220 feet ASL. It is on a 3000 sq.m. farm bounded by forage of forest trees in the south and by banana plants and coconut palms. A field of weeds, grasses and bushes is located on the North and East of the farm.

There are at least 150 colonies of *Tetragonula biroi* and six colonies of a clustered brood stingless bee of another species. Most colonies were sourced locally with at least 50 from other areas in Quezon. The farm was established four years ago with 10 locally sourced stingless bees.

The plants are well taken care of and the flower colour choice and composition planting are admirable.

Marvin Creencia Manila from Los Baños and Bernard Tuzon Anciado from Alaminos were there to join the crowd. We all then sat down for a spot of Blue Ternatea rice for lunch.

Figure 62The lovely entrance to Pia's Garden and the walkway to his Gazebo. The garden features a reception area of sorts with an upper terrace overlooking the garden.

After lunch, I was given a guided tour of Pia's Garden by Lee Gaitana himself. It had well-organized positioning of hives and flowering vegetation. I mentioned that he was lacking some flowering groundcovers, however, he assured me that he did have them previously but when he reared geese recently, they 'slurped' up all the grasses and flowering weeds including the groundcovers he prepared for the bees.

Figure 63 The Bee farm had much to offer in terms of a variety of species and hive structures.

Figure 64 Assorted Tetragonula Stingless Bee hives of T. biroi, T. laeviceps and T. sapiens

The group interactions were interesting with Bee Merry's Lipa Beekeepers coop members, Los Baños representative (Marvin Creencia Manila), Alaminos representative (Bernard Tuzon Anciado), Tayabas representative (Dr. PJ de Castro) and Mataas-na-Kahoy representative (Cecile Cudia) plus one Malaysian.

Figure 66 'Toping' style of meliponiculture is also practised amongst tiered box methods.

Figure 65 The group interactions were interesting with Lipa Beekeepers coop members, Los Baños representatives, Alaminos Rep., Tayabas Rep., and Mataas-na-Kahoy beekeepers plus one Malaysian.

Figure 67 The farm had hive boxes on racks and well shaded under magnificent trees and coconut palms

The farm had hive boxes on racks and was well-shaded under magnificent trees and coconut palms. Lee Gaitana has a Gazebo for social interactions with fellow beekeepers, built from a bamboo structure with woven bamboo screen rails.

5.1.1 Addendum - Pia's Bee Farm, Lipa City, Batangas by Lee Gaitana

Pia's Bee Farm, patterned after the Beescape concept, is owned and operated by M. Sgt. Lee Gaitana, a retired Air Force Enlisted Personnel.

It was named after his only daughter Pia who is growing up exposed to beekeeping. From her infant to toddler years, she spent most of her time on the farm or with his father during beekeeping consultancy visits. Pia is the namesake of Padre Pio, her patron saint.

During four years of operation, it has evolved from a simple apiary of native honeybees, *Apis cerana*, into a farm-to-table restaurant, training facility and a haven for native and exotic honeybees and stingless bees.

Image 6 The eatery by night

Image 5 The only daughter Pia

Image 7 Award two years after retirement for community involvement considering bees

5.1.2 Retirement Plan

Five years before his retirement, he made plans to keep bees as a hobby. He attended the UP-Beekeeping Training Program a few months after his separation from military service. There he learned about the stingless bees which were new to him having known only honeybees as the bees to keep. Upon his retirement, he bought a piece of land and kept *Apis cerana*.

5.1.3 *Apis cerana* Failed Venture

Apis cerana was the owner's first bee of choice. He found out later that the bees swarm and divide profusely such that in a year his initial 14 colonies rose to 150.

It was a good first year in a supposed hobby turned into a business. However, he realized that *Apis cerana*, sugar feeding, and selling sugar syrup honey, was not the thing he had in mind for keeping bees.
After just a full year, he gave up *Apis cerana*.

Image 8 Fully equipped and attired and ready to proceed

5.1.4 Meliponine Journey

In 1999, while he was patrolling the mountains of Mindanao, he rested upon a tree and there he noticed a protruding clay-like tube with insects coming in and out. They did not seem to be like bees to him, they were black and small. Their CAFGU (Citizen Armed Force Geographical Unit) guide narrated that they were bees and that the honey was used to treat stomach aches. This was his first encounter with the stingless bees.

Upon giving up *Apis cerana*, he turned to stingless bees. He recalled his notes and started doing research. The problem back then was finding a mentor. Stingless bee management is a totally different thing. Rico Omoyon, a notable local beekeeper, was his only mentor through online talks. But it was just bits and pieces of information.
To broaden his knowledge about stingless bees he bought books, hunted feral colonies and started keeping them and experimenting with their management.

5.1.5 From Bees to Food

In its three years of operation, Lee has set up the place by planting rare fruit-bearing trees and melliferous plants. When the trees started to bear fruit, they attracted not only bees but local visitors and tourists as well. Then one day, a visitor asked, do you have food?

Image 9 Some local dishes and the Blue Rice speciality set

This brought in the concept of a farm restaurant that people who do not like city noise can go to.

Image 10 Apis mellifera hive box with bottled honey above it; Hive frame with bees buzzing

It is an alternative to fast food, fancy restaurants and expensive coffee shops. Here the visitors can enjoy the greens, chirping birds and buzzing bees.

5.1.4.5 Apis mellifera

Yet something was lacking. True, visitors were amazed by stingless bees. Almost always there would be a visitor who was in awe that this little black insect is indeed honey-producing. But most of them also wanted to see real honeybees.

Lee then decided to keep *Apis mellifera*, not for its honey-producing capability but to complete the tour experience for each visitor wanting to see bees.

From two colonies he now has ten and he can sell mellifera honey from his own produce.

5.1.6 The focus

Having kept *Apis cerana*, *Apis mellifera* and stingless bees, Lee wanted to focus on a particular species, the *Tetragonula biroi*. Again, mentorship was an issue.

Fortunately, a senior stingless beekeeper who advanced in his field came along, Abu Hassan Jalil. He is accommodating and is guiding him in his vision to learn more in the field of keeping stingless bees and in producing hive-derivative products which are less known in the Philippines. Moreover, his cooperation with Sir Abu has led him to manipulate propolis and its amazing benefits for medicinal purposes.

5.1.7 The Future of Pia's Bee Farm

Image 11 Submission of honey sample for honey competition during the AAA 2020

Pia's Bee Farm will continually evolve. A hydroponic production facility for farm-to-table food will be constructed. More products from propolis will be sold in his shop, as well as honey which is unique to the farm's forage. It will continue to recreate its surroundings to be Instagram-worthy.

The owner also wishes that his farm be officially recognized as an Agricultural Training Institute accredited learning site for agriculture (LSA) to follow in the footsteps of the amazing Abu Hassan Jalil in sharing knowledge about keeping bees.

Pia's Bee Farm will continue to be open to visitors not only as a farm-to-table restaurant and a tourist destination but more as a place to inspire others to keep bees, not as a money-making venture but as a way to help nature and conserve these wonderful insects!

5.2 Cecile Cudia, Manggahan, Mataas-na-Kahoy, Batangas

Cecile Cudia is the sister of Chito Cudia whom I have befriended for some time on Facebook. I told him I was touring Batangas province, and he agreed that I could visit his farm which is Mataas-na-Kahoy and is managed by his father and sister Cecile.

I met Cecile when I was in STIARC, Lipa City and where she guided us to Pia's Garden in Tambo. From there we then followed her to Manggahan, the location of Cudia Farm.

Figure 68 We made our way through much vegetation and dipterocarp saplings and trees amidst an undulating plot. I saw the log hives with toping boxes well spaced apart.

Figure 69 They showed me a few trees which had some feral hives that naturally developed with abundant vegetation.

I met Cecile when I was in STIARC, Lipa City and where she guided us to Pia's Garden in Tambo. From there we then followed her to Manggahan, the location of Cudia Farm.

We made our way through much vegetation and dipterocarp saplings and trees amidst an undulating plot. I saw the log hives with toping boxes well-spaced apart.

The walk up and down the undulation was quite daunting for me but I managed with some stops to breathe and of course under the pretext of looking for feral hives in the trees.

5.3 Bee Merry (Cecilia Katigbak De Jesus) – Lipa City

From Manggahan we proceeded back to Lipa City to The Beehive Restaurant to view the Apiary in the backyard. We were invited to stay for dinner courtesy of Bee Merry, the Batangas Beekeeper Coop chief. It was interesting to see the menu was very much Apiary influenced and the names of dishes were mostly names of bees. After a photoshoot with Dr. PJ, he had to leave and could not join us for dinner.

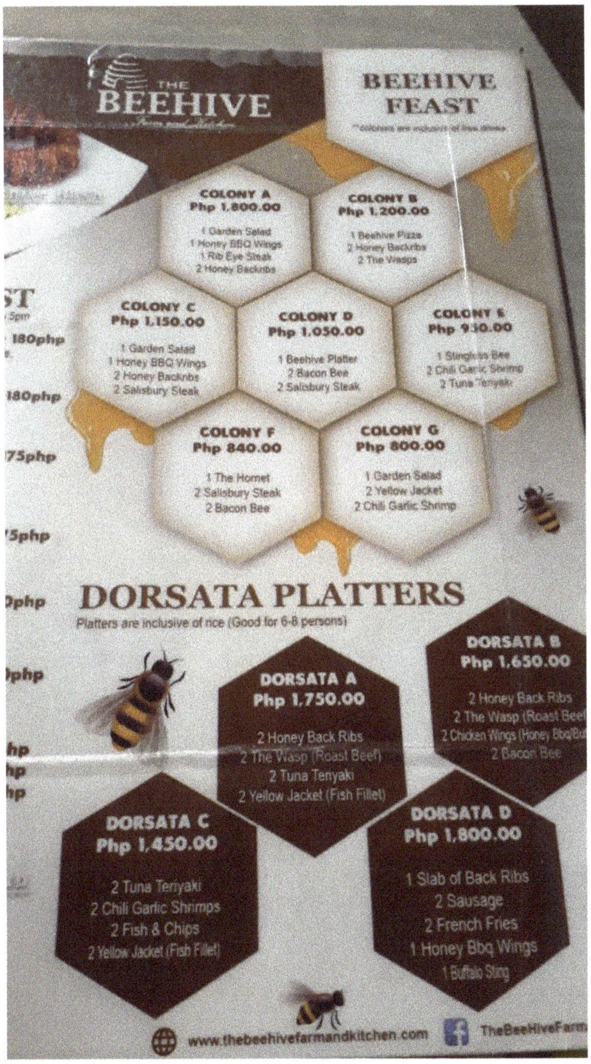

Figure 70 the menu was very much Apiary influenced and the names of dishes were mostly names of bees. After a photoshoot with Dr. PJ, he had to leave and could not join us for dinner.

A narrative summary from the Beehive Restaurant Facebook promotion video:

In 2008 someone let him taste the honey cider. He loves cooking and eating, so he decided to attend the seminar to learn how to make vinegar. The Beehive Farm and Kitchen... the house was built in 1998 and in 2015 the farm owner became a member and officer of the Lipa Beekeepers Marketing Cooperative. They held meetings in the said house and after the meeting, everybody was hungry and looking for food,

Figure 71 The house converted Restaurant with an elaborate entrance. The Apiary in the backyard had a few hive boxes of stingless bees.

so they decided to turn it into a restaurant. The difference is they have a farm and they don’t need to buy some of the ingredients in the market...like honey and vinegar. The best-seller on the menu is the baby back ribs. Many clients come to buy honey and if the clients want to have a tour and study about bees, how to take care of the bees and how the bees produce honey, then they teach them.

5.4 Calatagan Mango Farm Visit

5.4.1 Etymology: Calatagan. The word "Calatagan" is taken from the Tagalog word "latag" and is closely associated with "kapatagan", which means a vast portion of flat land lying between the hills and mountains. Thus, Calatagan means a large expanse of wide flat land.

5.4.2 Getting There: It takes about 1.5 hrs by road transport from Cavite the adjacent province. Calatagan is in Batangas on the west of Taal Lake and at the southwestern tip of Luzon. True to its name, Calatagan is a wide expanse of flat land. Ideal I guess, for mango farms.

At the front of this Mango farm is a pasture of flowering annuals and the farm consists of mango trees of various ages. The largest was 20 yr. old and some trees were only 2 yr. old.

Figure 72 Frontal portion of the mango farm with a flowering pasture

5.4.3 Flower pasture[4] at the farm entrance area

Figure 74 Hives are placed on metal stands in clusters of four and have a density ratio of 1 colony per tree.

Figure 73 The colonies used were of Tetragonula biroi and the hives had wire mesh at the entrance holes to prevent intruding pests from entering.

[4] A flower pasture in the making at the entrance area, although one would expect some beehive to be at farm fringe, closer to this pasture. This would be a welcome source for the bees in times of the off-mango flower season.

Figure 76 The canopy of the older trees was estimated to be about 150 m2. Mangoes are wrapped to protect them from fruit flies

Figure 75 Some stations had Bee 'hotels' to provide for solitary bees which may assist in pollination.

The older trees (20 yr.) were estimated to be 14 m apart, and the younger trees were about 5 m apart. It is not possible that the saplings were initially planted 14 m apart. Saplings would have been planted about

2.5 m apart, and then gradually thinned as they grew through the years. In total there are about 100 trees with stingless bee box hives in the same amount with additional solitary bee boxes in a few places.

To avoid older trees growing too tall the farm management team regularly does a yearly pruning of the apical branch so as to allow lateral growth, hence enlarging the canopy. The common pests are fruit flies and to prevent damage fruits are wrapped individually before ripening. We noticed some tall bamboo ladders here and there. The canopy size would call for some significant pest control measures and some form of pesticide must be strategically used at appropriate moments in order to not endanger the bees.

List of Figures in Part 1

Part 2 – MIMAROPA Tour

Abu Hassan Jalil, Nora Alanza Peña & Romblon. Photos: Joe Asis

Preliminaries

This is a continuation of the Meliponiculture tour series in the Philippines. The first of the series, the South Luzon Tours, covered the provinces in the CALABARZON Region and included Romblon Province. This tour reached the southwestern tip of Batangas and the south-eastern tip of the Bicol Region. The second of the series, the Visayas Meliponiculture Tours, covered six major islands of the Visayas. This third series sees the exploration of Mindoro and Palawan, which are the two major islands of the MIMAROPA Region.

This book includes excerpts of the main principles of Advanced Meliponiculture from a course conducted in Lipa City, Batangas for a score of Meliponiculturists convened from various parts of the Philippines.

Acknowledgements

The authors wish to thank the editors, Dr. David Ward Roubik for the technical issues and most of the English text and much gratitude to Danny Viola Hizon for some English portions and edits on all the Filipino text.

Much gratitude to Ms Reynalyn Pilar who provided us with much information and images on the Agricultural Cooperatives in Marinduque. Without her vital information, we would not have been able to chart them.

NAP acknowledges Rowena Alanza-Mararac (NAP's sister) for her geniality and accommodation in Sta. Rosa, Laguna, Alan Cabalfin for all his advice and guidance, and Lee Gaitana and family for their hospitality in Lipa City, Batangas. Many thanks to her daughter Bethel Joy Peña for refining the English text in the diaries portion and the overall book cover design.

We express our gratitude to all the Stingless Beekeepers, Meliponary and farm owners that shared their hospitableness during our Tour in MIMAROPA. Much gratitude to all farm owners who allowed us to take photographs of their Meliponary setup, stingless beekeeping activities and gardens.

Special thanks to Victorino Samarita and Mar Bautista for their services as guides and Neth Pastor for the use of her vehicle throughout Oriental Mindoro. We are grateful to Ric Aranas (a colleague of Victorino)

for allowing us to be accommodated in his house in Barangay Hagan, Bongabong Oriental, Mindoro and to Donato S. Cuden with his colleagues in guiding us throughout Occidental Mindoro. Without all their assistance, the round Mindoro Tour would not have been possible.

We thank Jun Tabinga and Joseph Dador for aiding and guiding us throughout the Palawan part of the tour and without them, the Palawan trip would not have been successful.

We acknowledge Neth Pastor, Perry Mateo, Albert Mangapit, Perfecto Diaz, Danny Hizon, Lelet Tabinga-Stolz, Keith Mikkelson, Lee Gaitana and Leo Grajo who contributed profile articles and photographs of their farms to furbish the manuscript.

Article Contributors

		Page
1	**Albert Mangapit** - *Faith Paradise Agri-Tourism Farm - How the farm came about*	131
2	**Danny Hizon** - *Profile: Danizon Farms, Sablayan, Occidental Mindoro*	185
3	**Keith Mikkelson** - *Stingless bee colony photos*	230
4	**Lelet Tabinga-Stolz** – *Profile: Haley's Bee Farm, Puerto Princessa, Palawan*	217
5	**Neth Pastor** - *Profile: Ramona Mantaring Pastor & HN Organic Farm*	161 - 163
6	**Perfecto Diaz** – *Profile: DosfourDos Integrated Farm*	243
7	**Perry Mateo** – *Profile: Mateo Meliponary, Pinamalayan, Oriental Mindoro*	*140*
8	**Reynalyn Pilar** - *Meliponiculture activities in Marinduque*	259
9	**Rodolfo O. Abalus, Jr** – *Stingless Bee Program, Yamang Bukid Farm, Palawan*	226

Foreword to Part 2

No-sting in a tale.

A few months ago, we embarked together on a mission to come to the Philippine Islands and discuss some fairly important issues in our realm of expertise. I confess to liking ecotours and also reading, so, logically, having the chance to participate in the former and then read about it was absolutely great. Thanks to the people who appear through the pages of this publication, we can all live or relive the many fascinating and pleasurable days which were spent getting to know our subject, and each other, quite a bit better.

The highlight of the trip was learning how an individual, frustrated with honeybee keeping, made a revolutionary observation of how stingless bee colonies could be kept in coconuts, after taking down nests that were on the trees themselves. In contrast, the main disappointment was finding that, although such stingless bee colonies are employed for pollination in orchards, the pesticide application continues to disrupt pollination management and kill the bees. We can all do much better than that.

Our travels, so richly detailed here, are part of an ongoing story, and considerations of the quality of life. I hope that its readers will feel better, as I do, by finding for themselves what we have been privileged to find here.

David Ward Roubik,
Senior Scientist,
Smithsonian Tropical Research Institute,
The Republic of Panama.

Preface to Part 2

The whole tour is divided into five parts. The first leg of the tour covered attending the Asian Apiculture Association (AAA 2020), the 2nd International Meliponine Conference and BEES 2020. These events were held at the University of Philippines, Los Baños (UPLB) in Laguna Province, Luzon from 25th Feb to 29th Feb 2020. In this leg of the tour, the organizers included a technical workshop in Mabitac, Laguna, a visit to Ilog Maria Honeybee Farm in Silang, Cavite Province and a Mango farm in Calatagan, Batangas Province.

On the 2nd leg, we were flown to Legazpi City, province of Albay, Bicol Region to visit Balay Buhay sa Uma Bee Farm in Bulusan, Sorsogon Province. Here the learning covered collecting a fern cluster nest and hiving it and splitting a year-old coconut kernel cluster hive. We were in Sorsogon from 1st March to 3rd March and we had the opportunity to visit some places of historical interest and some tourist spots.

The 3rd leg then saw us in Lipa City, Batangas province attending an Advanced Meliponiculture course together with beekeepers from various parts of the Philippines excerpts of the main principles of advanced Meliponiculture are included herein.

The 4th leg of the tour is the part where the MIMAROPA tour starts. This portion of the book starts with the ferry ride to Calapan City, Oriental Mindoro. The trip practically took us around the whole of Mindoro Island for over two days. Leaving Mindoro from Abra de Ilog port, we proceeded to Ninoy Aquino International Airport via Batangas to catch the early morning flight to Puerto Princessa.

The final portion, the 5th leg, saw the tour of Palawan Island over another two days visiting bee farms and places of interest. This whole tour and book publication were done during the COVID -19 Pandemic lockdown period i.e., Feb to May 2020.

Having scoured through many locations, I developed a casual format. First, I would map the location and explore the modes of getting there from an entry gateway like an airport or port terminal or the previous destination. Next would be the etymology or the history of the place. Then an overview of the peoples, cultures and ethnicity. Then we would preview the local flora and fauna especially if there are any peculiarities. After all these, we would look at the bee diversity and the endemic stingless bees. The meliponiculture would follow with a focus on regional methods and various styles of obtaining the products from the stingless bees. The beekeepers themselves will be the highlight eventually.

Introduction to Part 2

This publication was prompted by the Asian Apicultural Association Philippines Symposium on Pollinator Conversation - AAA 2020 and the 2nd Meliponiculture International Conference and the BEES 2020 events, all held at the University of the Philippines Los Baños (UPLB) from 25th to 29th Feb 2020. the event organizers, as part of the activities for the invited speakers, arranged for a workshop on Meliponiculture held at the Mabitac Kota Paradiso farm, and an excursion to Ilog Maria Honey Bee Farm in Silang Cavite, then a Mango farm to observe first-hand the pollination activities by Stingless Bees in Calatagan, Batangas. The next day we were flown to Legazpi City in Albay Province to visit Balay Buhay sa Uma farm in San Roque, Bulusan, Sorsogon Province.

After two days in Sorsogon, we came back to Manila and travelled to Lipa City in Batangas Province to attend an advanced meliponiculture course. When the course was completed, we proceeded to the region of MIMAROPA.

MIMAROPA means the provinces of Mindoro (M. Oriental and M. Occidental), Marinduque, Romblon, and Palawan)

The MIMAROPA Region, officially known as the Southwestern Tagalog Region, is an administrative region in the Philippines. It was also formerly designated as Region IV-B until 2016. It is one of two regions in the country having no land border with another region (the other being Eastern Visayas). The name is an acronym combination of its constituent provinces abovementioned.

Mindoro (Occidental and Oriental) - Hispanicized form of minolo or mintolo, a local word of forgotten meaning, referring to the name of Mindoro Island's principal trading town during the early Spanish colonial period. (See Fourth Leg Map 7 - Fourth Leg Mar 6th – 7th 2020)

Marinduque - Hispanicized corruption of malindig or malindug, Tagalog for “tall and elegant”, referring to the tallest peak on the island.

Romblon - Early Spanish accounts rendered the toponym as Donblon in Spanish orthography, which is probably based on the native word lomlom, a term with cognates across many Philippine languages meaning “dark”, or “shady”, perhaps a reference to the once thick forests of the place or the clouds that constantly form over it.

Palawan - Palawano form of perawan, Malay for "virgin land," about the largely untouched resources of the main island that now bears its name. (See Fifth Leg)

Reference: https://en.wikipedia.org/wiki/List_of_Philippine_provincial_name_etymologies

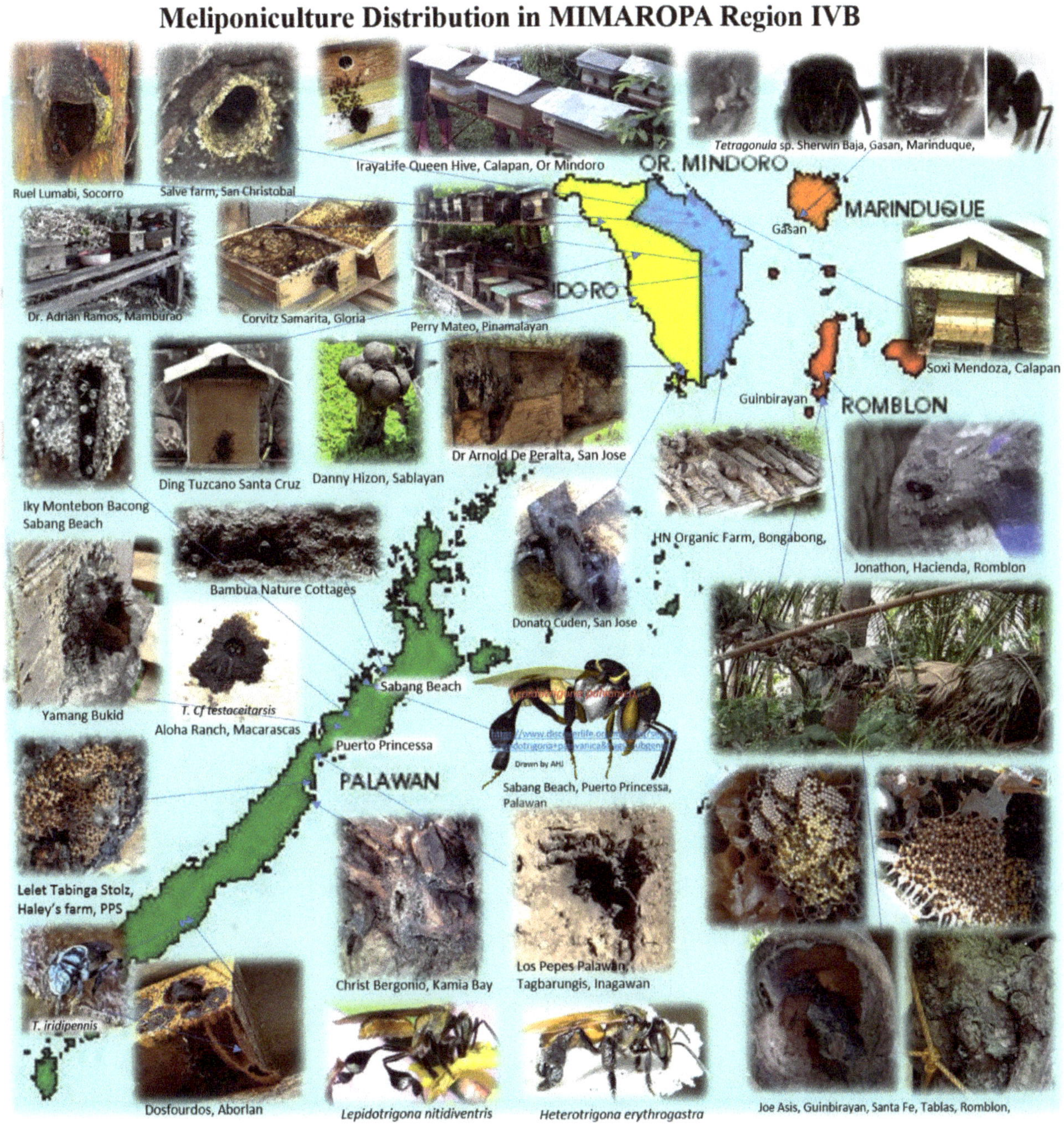

Chart 3 Chart of Meliponiculture activities in MIMAROPA Region IVB

List of Meliponiculture activities in MIMAROPA[5] Region IVB

1. Iraya Life Queen Hive, Calapan, Or Mindoro
2. *Tetragonula* sp. Sherwin Baja, Gasan, Marinduque,
3. Ruel Lumabi & Albert Mangapit, Socorro, Naujan
4. Salvacion Mariano, Salve farm, San Christobal, Victoria, Or Mindoro
5. Victorino Samarita, Gloria, Or Mindoro
6. Perry Mateo, Sabang, Pinamalayan, Or Mindoro
7. Soxi Mendoza, Calapan, Or Mindoro
8. Iky Montebon Bacong Cabayagan. Sabang Beach, Palawan
9. Fernando Ding Tuzcano Santa Cruz, Occ. Mindoro
10. Danny Hizon, Sablayan, Occ. Mindoro
11. Dr. Arnold De Peralta, OMNC, Camwali, San Jose, Or Mindoro
12. Neth Pastor, HN Organic Farm, Bongabong, Or Mindoro
13. Jonathon, Hacienda, Romblon
14. Joe Asis, Guinbirayan, Tablas, Romblon,
15. Kevin Dartman, Bambua Nature Cottages, Cabayagan, Palawan
16. Donato Cuden, & Zeny Topado, Bubog, San Jose, Occ. Mindoro
17. Yamang Bukid, Puerto Princessa, Palawan
18. Aloha Ranch, Macarascas, Palawan
19. Lelet Tabinga Stolz, Haley's Farm, Puerto Princessa, Palawan
20. Christ Bergonio, Kamia Bay & Los Pepes Palawan, Tagbarungis, Inagawanm Palawan
21. *T. iridipennis* Perfecto Dias, Dosfourdos, Aborlan, Palawan
22. Dr. Nathaniel Dezcano & Adrian Ramos, Jopson Farm Brgy.8, Mamburao, Occ Mindoro.
23. Joseph Dador, Brgy Irawan Puerto Princesa City Palawan

---

[5] MIMAROPA Region IVB
Romblon which is a part of region IV-B (MIMAROPA). MIMAROPA means MIndoro (M. Oriental and M. Occidental, MArinduque, ROmblon, PAlawan
The MIMAROPA Region, officially known as the Southwestern Tagalog Region,[3] is an administrative region in the Philippines. It was also formerly designated as Region IV-B until 2016. It is one of two regions in the country having no land border with another region (the other being Eastern Visayas). The name is an acronym combination of its constituent provinces: Mindoro (divided into Occidental Mindoro and Oriental Mindoro), Marinduque, Romblon and Palawan.

Chapter 6 Romblon Province

6.0 Tugdan Airport, Tablas, Romblon

Leaving Luzon from Manila, I flew south to Tablas Island, Romblon Province. The airport Tugdan is sufficient to receive small commercial passenger aircraft. Joe Asis was kind enough to fetch me from the airport. He was there with the Fire Chief of Alcantara, SF02 July B. Teologo.

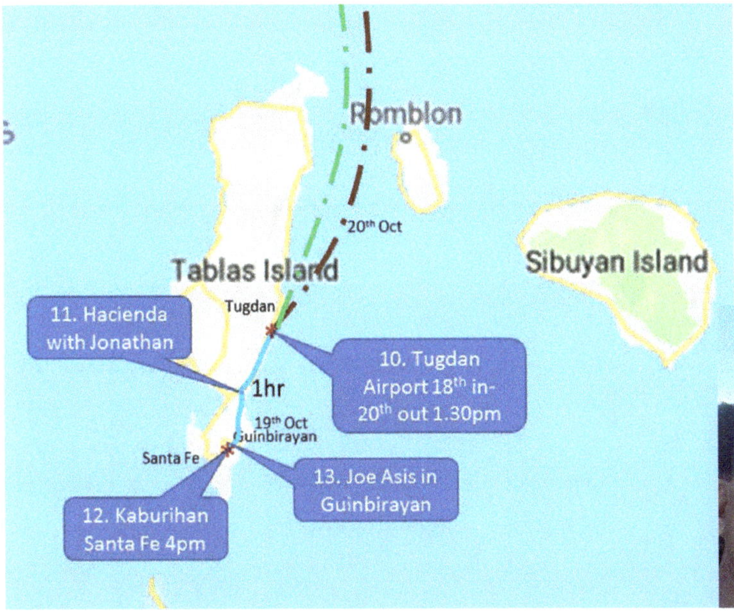

Figure 77 Leaving Luzon from manila, I flew south to Tablas Island, Romblon Province. The airport Tugdan is sufficient to receive small commercial passenger aircraft. Joe Asis was kind enough to fetch me from the airport

The one-hour flight from Manila was pleasant and it saves me from the 8-hr. ferry ride from Batangas to Odiongan, as originally planned. Joe fetched me in a gleaming red 4WD, a typical colour choice of the firefighting discipline, I thought. It belonged to one of his station colleagues. The first stop was the Alcantara fire brigade station for a spot of coffee with the fire Chief and staff.

I was intrigued by a mango tree beside the fire station, a mango tree that the fire Chief estimated to be at least 80 yrs. old. It looked to me almost 100yrs old at more than a 1-meter diameter. Joe said there is another one like it near his house. The photographs show the extent of the trunk diameters.

Figure 78 The humongous mango tree at the Alcantara Fire Station and the other slightly smaller one at Joe's house in Guinbirayan.

So, with that, we headed for Guinbirayan and along the way we stopped at a place Joe calls Hacienda (I presume from the word Hacienda) to meet with his hermit-like recluse friend Jonathan. - Hacienda

The spot along the unfinished highway was difficult to spot with no signage or landmarks. One had to traverse down on foot some 300 meters or so, and the elevation was roughly about 80 meters. Halfway down the valley, we came by a bamboo hut and Joe called out Jonathan's name a few times, but no reply. That is where Joe showed me what he called a Lib-og feral hive in a tree over the hut, I zoomed in and took a shot and estimated it to be about 5 to 6 meters high. Joe told me this species is different and a bit larger than the monggo monggo species.

6.1 Jonathan. - Hacienda

Joe brought me to meet with his hermit-like recluse friend Jonathan

The spot along the unfinished highway was difficult to spot with no signage or landmarks. One had to traverse down on foot some 300 meters or so, and the elevation was roughly about 80 meters. Halfway down the valley, we came by a bamboo hut and Joe called out Jonathan's name a few times, but no reply. That is where Joe showed me what he called a Lib-og feral hive in a tree over the hut, I zoomed in and took a shot and estimated it to be about 5 to 6 meters high. Joe told me this species is different and a bit larger than the monggo monggo species.

Figure 79 A scenic view from Jonathan's wilderness is a valley down about 80 meters from the unfinished constructed Guinbirayan highway from Tugdan

Jonathan showed us a spot where the soil has eroded and exposed a portion under the taproot of a big tree. At the buttress, there were 2 small nest entrances and one was about 5 cm long, looking to be similar to a *Tetragonula sapiens* entrance. After taking pictures and having a bit of a casual chat we bid Jonathan farewell and climbed back up to the highway.

Figure 80 Lib-og feral hive in a tree over the hut, I zoomed in and took a shot and estimated it to be about 5 to 6 meters high.

6.2 Kaburihan Beach Resort

We arrived in Guinbirayan village and my first thought was to check in first and have a rest after a full day of travelling.

.

Figure 81 The Kaburihan Beach Resort

Joe brought me to the Kaburihan Beach Resort. It is a very quaint village cottage kind of resort. Homely and cosy, it has all the necessary basic amenities for a restful vacation.

At the Bayshore, I could see the ridge that I photographed from the highway side at the spot of Hacienda where I visited Joe's friend Jonathan.

The next morning after sunrise, the resort owner, knowing that I was there to look at stingless bees told me that they discovered some broken branches that had some beehives, after a storm. They asked me to look at them and asked me to help in advising them on how to culture them. I obliged and upon inspection, found that the broken branch fell due to the rotting wood caused by termites and ants.

Figure 82 Most of the SB nests were at the nodes of the branches.

Figure 83 Fallen branches found behind the Kaburihan beach resort

It seems in that area, stingless beehives practically drop from the sky after a storm. The branches break off easily due to rotting and the weight of the nests within. In some, I found ant nests in the hollows and even termites in the deadwood living together with the bees.

Behind the Resort is the village of Guinbirayan amidst some secondary bushes and wild trees. Most of the SB nests were at the nodes of the branches. One was in a rotted coconut trunk and was on the roof of their chicken coop.

The hives were quite established in the hollow of the branches and cohabitating with the ants and termites. Some branches were infested with cockroaches.

Figure 84 One was in a rotted coconut trunk and was on the roof of their chicken coop.

Figure 86 The hives were quite established in the hollow of the branches

Figure 85 Cockroaches abound under the dried bark and remnants of some ant nests.

Figure 87 Managed to extract 1 tumbler full of wild honey from 4 broken branch hives.

With the help of Joe and Nora, we managed to extract one tumbler full of wild honey from four broken branch hives. Then the brood of each was put into split bamboo and tightly secured with tire tube rubber strips, one was capped with a coconut shell to be added to the previous entrance cerumen, and the one in the partial coconut tree trunk was put in one of Joe's boxes.

The bee colonies took to their new accommodation readily and happily as they settled into each new 'home'. It was quite satisfying for the day as we saved the colonies from inevitable despair.

6.3 Joe Asis - Guinbirayan

From Kaburihan, Joe led us on a morning walk to his farm in the village. I took pictures of groundcovers and low shrubs and asked for the local names of the plants. This was in preparation for an impromptu 'lecture' Joe requested me to do with his neighbours and friends in the community. I just instinctively knew the bee farm would not particularly have them for the bees alone.

Figure 88 On arrival we met Joe's wife with Pastor Roland Forcado and Fire Chief SF02 July B. Teologo was also there and Joe's wife prepared Brown rice with crabs and sweet potatoes.

Assorted flowering groundcovers and shrubs include Thunbergia, Wedelia, *Mimosa pudica* (Makahiya), Angelonia, *Oxalis berbieri, Merremia peltata,* Sweet Potato (Kamote), *Melastoma* sp. (Busisi). True enough the farm we arrived at had flowering trees, coconut palms and some skeletal shrubs. Lacking in low shrubs and flowering ground covers, I explained to Joe why I was taking pictures of the wild weeds along the way. On arrival, we met Joe's wife, Doreen with Pastor Roland Forcado and Fire Chief SF02 July B. Teologo was also there. Joe has 5 children all boys ages from youngest at 4 yr. to eldest at 12 yr.

Figure 89 Joe uses coconut husks as hanging planters

6.3.1 At Guinbirayan village

After viewing his hives, I was asked to provide some tips and advice on Meliponiculture. Then they even prepared a blackboard and chalk for me to conduct an impromptu class under the shade of the trees!

Figure 90 A peculiar nest contained in a plastic tub

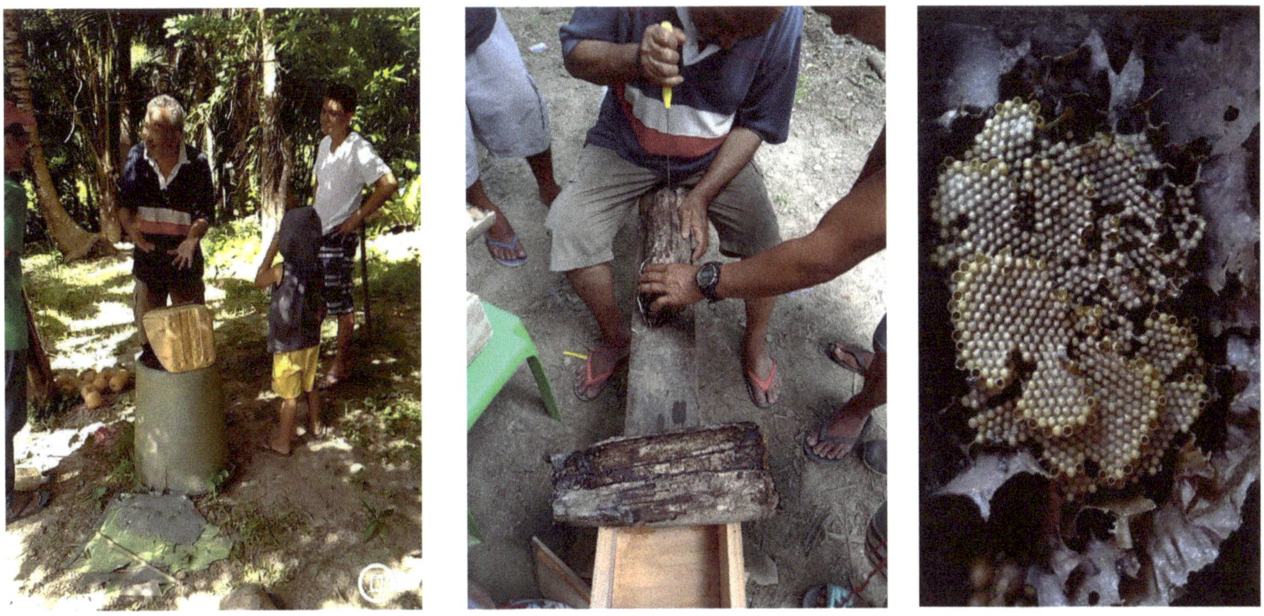

Figure 91 Inspecting a hive which Joe says the local name is Lob-og

Figure 93 The layered brood combs of Lob-og

Figure 92 Bamboo reinforced hives in rotting branches, as what we saw behind the resort. These, however, were not relocated yet but reinforced temporarily.

Figure 94 Other monggo monggo hives possibly T. laeviceps and T. sapiens

The videos of the workshop can be viewed at

https://www.facebook.com/abu.h.jalil.1/videos/pcb.1357232094443326/10220073711623925/?type=3&theater&ifg=1

Chapter 7 Oriental Mindoro

Map 1 Routes of the Mar 2020 Mindoro tour

7.1 Mindoro Tour Initiate – Batangas Port

7.1.1 Getting there: - Batangas Port. Upon completion of the Advanced Meliponiculture Course (see Appendix) in the evening, we said our goodbyes to all the participants and our host Lee Gaitana gave us a lift to Batangas Port to start our Mindoro tour. We were pre-booked at The Wharf Residence Hotel nearest to the ferry terminal. That night Victorino Samarita had come over from Mindoro to fetch us and to make sure we boarded the right ferry. I guess his function as the official Mindoro Guide started then. Our destination was Calapan Port in Oriental Mindoro. We left the next morning to catch the 6 a.m. ferry to Calapan in Oriental Mindoro.

7.1.2 Etymology: Batangas Port. The modern name "Batangas" is derived from Spanish batangas, meaning "outrigger [booms]". It is derived from Tagalog batangan which refers to a type of tough mangrove wood used in building the large rounded beams (also called batangan) across the traditional outriggers (katig) of Filipino bangka boats.

7.2 Calapan Port:

7.2.1 Getting there: We arrived after a 3-hour trip and were met by Neth Pastor and her chauffeur Mar Bautista and the Parking lot. All loaded up with the help of porters the vehicle's back portion was filled up. The thing was we were not just backpacking, NAP, Dr. Roubik and myself had luggage enough for 2 weeks because we started more than a week ago in Laguna.

7.2.2 Etymology: Calapan. There is speculation it came from the word "Kalap" which means to gather logs. Thus "Kalapan" was supposed to be a place where logs were gathered. In the old records, however, there was never a mention of Calapan as a place where logs were produced or exported. Furthermore, huge forest trees where logs were produced certainly did not grow near the town, which was swampy. Another theory holds that Calapan was originally pronounced as "Kalapang" which, according to an old Tagalog dictionary, was a synonym for "sanga" or branch. It could then refer to the settlement of Kalapang as a branch of its mother town of Baco, an adjoining town. The name was later hispanized as Calapan

7,3 Salvacion Mariano ng Brgy San Cristobal, Victoria, Oriental Mindoro

We arrived at Mam Salvacion Mariano's Farm at 8:43 a.m. She let her son meet us at the corner street and guide us to her farm. They have a large farm and they just started landscaping. With Mark's help, we were able to locate the bee colony in a dead tree. Dr. Roubik got some specimens and while he was, I watched his technique of catching bees. Some of the bees bit us because Dr. Roubik peeled and removed the portion of their nest entrance.

Mrs. Mariano brought us to a nearby farm named Norma's Bread and Breakfast. According to her, the farm is three hectares and is home to different fruit trees. We checked the bee nest and took photos. After a few minutes, we gave thanks and bid goodbye.

7.3.1 Salve Farm, San Cristobal, Victoria:

Figure 95 Arrival at Salve farm and a group photo at the Main gate, from left Mark Daniel Mariano and his mother Ms. Salvacion Mariano, AHJ, DWR, NAP, Neth Pastor & Victor Samarita

7.3.2 From Calapan Port Visiting Ms Salvacion Mariano & her son Mark Daniel Mariano of Salve Farm in San Cristobal, Victoria, Or. Mindoro, we made our way through Naujan District to San Cristobal. On a dirt road, Mark Daniel Mariano guided us to reverse the vehicle because there was no space to turn up ahead. We soon found ourselves at the front of a plot that was being worked on to be Salve Farm.

Figure 96 There were a couple of rest huts made of bamboo and we found quite a number of nest entrances built in the joints and the bamboo structures themselves

7.4 Visiting Norma B&B, San Cristobal, Victoria, Or. Mindoro.

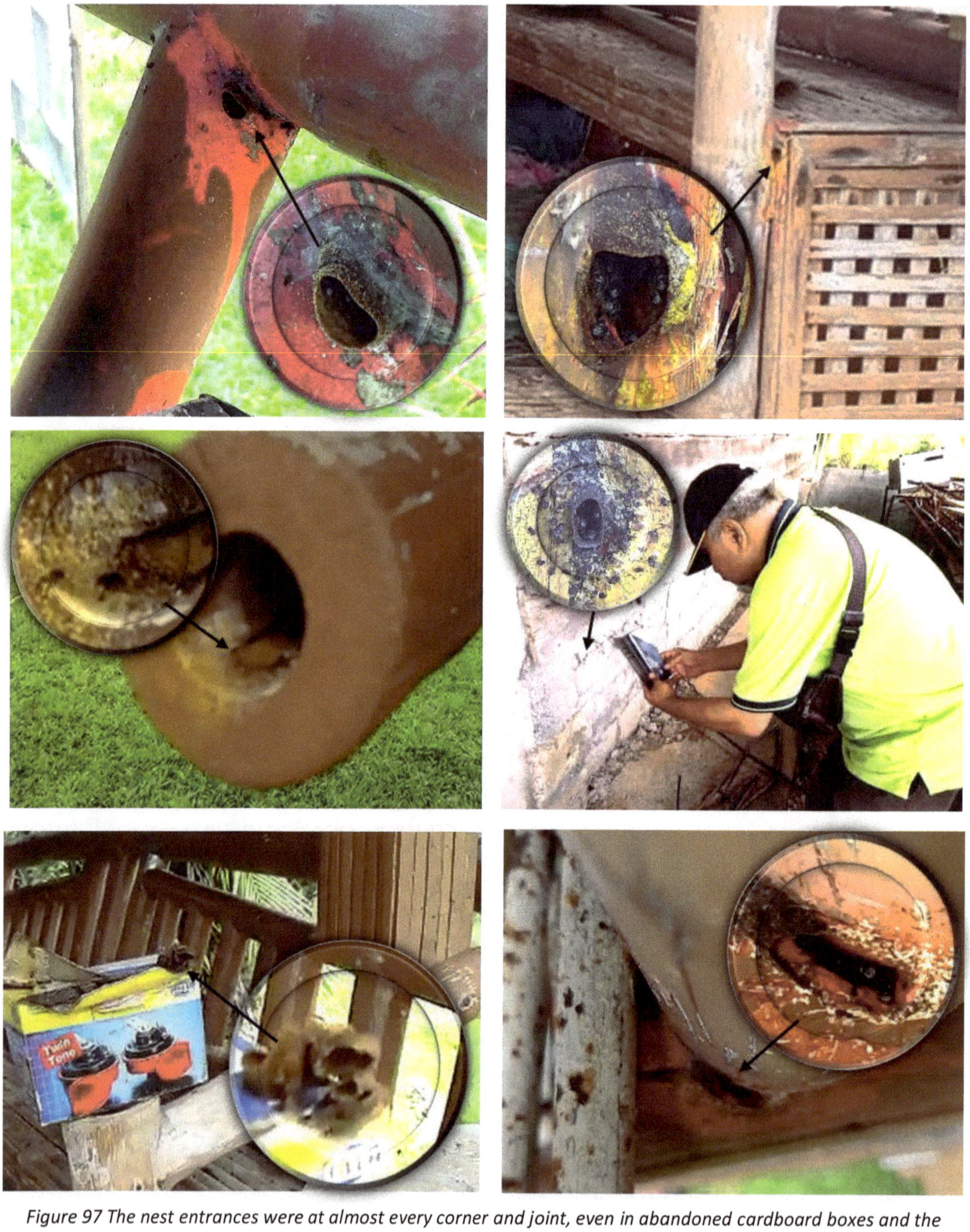

Figure 97 The nest entrances were at almost every corner and joint, even in abandoned cardboard boxes and the foundation cracks and walls

7.4.1 Etymology: San Cristobal – Named after the town's patron saint, Saint Christopher.

Figure 98 A couple of entrances adjacent to one another at the pipe and drainage elbows

7.4.2 Feral colonies[6]: Such is the situation in which almost any corner or opportune cavity is occupied by a SB nest. In abandoned cardboard boxes and the foundation cracks and walls; in the bamboo structure and joints, they seem to enjoy reproducing and making nests in every conceivable spot. Even in pipe elbows and drainage fixtures as can be seen in the photo above.

[6] There were no domesticated bee colonies on this location but were abound with feral colonies in the bamboo furniture, bamboo structures and in the walls and foundation cracks. See previous page.

7.5 Albert Mangapit, Faith Paradise Agri Tourism Farm, Socorro, Oriental Mindoro

7.5.2 Getting there: A short drive of about 45 minutes from San Cristobal. We were greeted by Ruel Lumabi the caretaker of the farm

7.5.3 Etymology: Socorro The name **Socorro** is a girl's name of Spanish **origin** meaning "succour, help, relief". **Socorro** is a Spanish word name taken from the title of the Virgin Mary, "Mary of Perpetual Succour." Its religious meaning makes it somewhat old-fashioned

Figure 99 From left Victor Samarita, NAP, AHJ, DWR, Mar Bautista, Neth Pastor, Albert Mangapit & Ruel Lumabi at the farm entrance

7.5.4 Agro Tourism Farm Our initial reception was greeted with an array of colourful flower beds. It seems to me that this farm is well-planned and designed for Agri Tourism. The entrance area is adorned with assorted flowering annuals and local shrubs.

Among the annuals are *Tagetes, Zephyranthes, Zinnia, Gomphrena, Portulaca, Celosia* and many more. The owners also prepared new flower beds to keep the continuity of blooming annuals.

After appreciating the local flowers, we were treated to some local dishes. It was a porridge-like dish with yam in coconut milk.

Besides wooden box hives, they also have coco kernel hives and these are hung below the racks. All these hives appear healthy and active. The farm is in an area of 2.5 hectares.

Figure 100 Assorted array of colourful flowerbeds

7.5.5 Eduction: Interestingly they experimented with some eduction efforts and have been successful thus far. The intriguing part is that they build the rack beside the tree that had a feral colony and attached the tube from that to the box hive placed on the rack. When that box has taken a new colony, the tube is detached and moved aside, and a new setup is put in place. One can simply go on filling up the rack with boxes, repeating the process in this manner.

For those boxes with the new colonies, one can even add a topping box to extend that new colony for honey harvesting without disturbing the 'satellite' colony. Besides, they have added a flower bed just below the rack to assure foraging sources for the new satellite colonies.

Figure 101 New beds are being prepared to assure continuity of blooming annuals consistently for foraging for the stingless bees

Figure 102 After appreciating the local flowers, we were treated to some local dishes. It was a porridge-like dish yams in coconut milk.

Figure 104 Besides wooden box hives, they also have coco kernel hives and these are hung

7.5.6 Addendum - Faith Paradise Agri-Tourism Farm (FPATF)

7.5.6.1 How the farm came about; by Albert Mangapit

Albert Mangapit and his wife Imee had been dreaming of having a farm. Although Albert had agricultural properties in Ilocos, the couple dreamt of a farm elsewhere. The urge of having a farm suddenly came during the trying times of their family. They wanted a sanctuary. While Imee had colon cancer then and while the couple were focusing on healing, one of their sons got involved in a serious problem. Imee opted for alternative healing to deal with her condition and they also saw the need to have a farm advocating organic farming for people to be healthy. Their question then was, where will they find a beautiful farm that is within their budget? Pastor Florencio "Boya" Ramos and his wife Solia, retired church mates from Faith Fundamental Baptist Church in Antipolo who are now living in Mindoro, invited them to visit Socorro and see prospective farm lots.

The couple was happy that they found the farm of their dream that same day, a place full of different fruit-bearing trees located just a few meters from the main highway. It was an ideal farm for the couple. Albert secured the place and then the couple prayed for a trustworthy caretaker who would look after it. A few months later, God answered their prayer and provided them with their church mates also from Antipolo, Ruel and Jo Lumabi. Albert and Imee are so blessed to have them because they are of the same faith and they love farming.

It was Imee who thought of the name. Why "Faith" and why "Paradise"? They believe God has been Faithful in their lives despite the trials their family went through, and Paradise because this is God's promise to those who are trusting Him. In honour of His goodness, Imee wanted a prayer garden on the farm where people can commune with our Creator. She visualised it with many colourful flowers and butterflies flying around, but she never thought about bees.

7.5.6.2 STINGLESS BEE

Image 12 Albert and Imee at the left while Ruel and Jo at the right side and Jonjon, Nicole and Ruelyn their children

Image 13 Visitors from Agricultural Training Institute during the inspection for accreditation of Faith Paradise Agri Tourism Farm

One morning, Albert was listening to a radio program about the care of fruit-bearing trees. Imee overheard the speaker, that to have an increase in fruit harvest there should be pollinators like bees and butterflies. It made her happy to know that having many flowers on the farm is beneficial. Between bees and butterflies, the owners decided to have stingless bees instead of butterflies because of the benefits of their honey to health and beauty.

Ruel found existing colonies of stingless bees in FPATF and he introduced eduction upon learning about it through the Agricultural Training Institute (ATI). Albert also obtained more colonies from his friends. At present FPATF has more than 30 colonies and it will have more in the coming years, by the grace of God.

7.5.6.3 FOOD SECURITY

PATF has a lot of Lanzones, newly planted Durian of different varieties, Strawberries, Mangoes, Papaya, Banana, Cactus Dragon Fruit, Peruvian Apple and Opuntia, lots and lots of flowers, root crops, herbs, vegetables and free-range chicken.

7.5.6.4 SERVICES

FPATF seeks to serve the community through its accreditation with the Agricultural Training Institute (ATI). Its role is to share knowledge by conducting training and seminars for farmers to increase their productivity and encourage the practice of organic farming. It can also serve as a sanctuary for prayer and peace with our Creator and as an educational touring place for students as soon as the required facilities are established.

FPATF also sells online via our FB page Faith Paradise Agri-Tourism Farm. Further, it is also the desire of FPATF to serve the people of Mindoro.

To God be the glory alone!

Image 14 FPATF has a lot of Lanzones, newly planted Durians of different varieties, Strawberries, Mangoes, Papaya, Banana, Cactus Dragon Fruit, Peruvian Apple, Opuntia, lots of flowers, root crops, herbs, vegetables and free-range chicken.

7.5.7 Naujan Lake, Oriental Mindoro is touted as the country's emerging eco-tourism destination. Albert Mangapit thus brought us to a location owned by Pastor Florencio Ramos, where we could appreciate a view of Lake Naujan. Naujan Lake is a freshwater lake located in the north-eastern corner of the province of Oriental Mindoro on Mindoro Island. The lake is the fifth largest in the country and the main geographical feature of the Naujan Lake National Park. The lake is also bounded by the Naujan Mountains to the north and east. Mount Naujan, at an elevation of 1,380 feet (420 m) one of the 70 inactive volcanoes of the country, is situated northeast of the geographic centre of the lake.

Figure 106 Top left: Lake view with Mt Naujan in the background; Top right: some nest entrances at the sidewall of the stairway; Bottom left: Group photo at the Lake view deck. Bottom right: NAP with Albert Mangapit

7.5.7.1 Etymology: Naujan, an earlier civilization existed, as evidenced by the unearthed artefacts of Chinese origin dating from the Sung, Yuan, and Ming dynasties in the barangays of Dao and San Jose. This reinforces the theory that the natives of Naujan were the Chinese during those early years. Its name according to legend came from the word, "nauhaw", or "went thirsty".

7.6 Perry Mateo, Brgy Sabang, Pinamalayan, Oriental Mindoro

11:53 a.m. Perry waited for us in the town centre to lead us to go to their barangay. While we were beside the river waiting for his father-in-law to pick us up, we took several photos of the river. Later on, tatay Gavino Ermita arrived. Inside the jeep, we looked for a place to sit. We made sure we had a strong grip on the side of the jeep. Kuya Mar was settled in the hammock. I was focused on the speed of the vehicle, so I failed to count how many rivers we crossed.

I was amazed at what I had seen at Tatay's farm. He has numerous colonies of bees. Each group has a separate housing facility. A big blue jar containing 50 kilos of Apis breviligula honey and a small one with around 25 kilos of Tetragonula biroi honey were displayed on the terrace. He has a huge parcel of land which is around 60 hectares. Nanay Nida, his wife, and Perry's wife Maureen prepared lunch for us. Also present was grandpa Melecio who is more than 90 years old. It was my first time eating steamed young fern that they called "kinilaw na paku". After lunch, we toured the stingless bee houses and the black pepper farm. We could not resist asking for some plants to bring home. Tatay Gavino was so kind to us. We soon said goodbye to the whole family as we still have other farms to visit.

While we were seated, more comfortably this time, inside tatay Gavino's jeep, I had the opportunity to count the number of river crossings we made. Well, only 10. But because tatay Gavino's jeep is big the crossings were not much of a problem. Before we broke up, we had a group picture with his jeep as our background.

7.6.1 Visiting Perry Mateo, Brgy Sabang, Pinamalayan, Oriental Mindoro

7.6.2 Getting there: The Barangay Sabang is about 20 km from Pinamalayan town. The town of Ipinamalay (now Pinamalayan) is close to the shore of the river of the same name about 2 and a half miles southwest of Bugot Point. The Pinamalayan River whose source originates from the mountains north of the town empties at the mouth of the old Lumang-Bayan River. It took about 1 hr to drive from Sabang to Socorro.

Figure 107 Side of Perry's House

7.6.3 Etymology: Pinamalayan has its history way back to Spanish colonization. Translated, Pinamalayan means illuminated, before this its name Ipinamalay meant 'release' in English.

7.6.4 Meliponary of Perry Mateo[7]

Figure 108 Traditional bamboo house with Perry's Meliponary in Sabang barangay

[7] Perry Mateo's father in-law Gavino Ermita, maintains another Meliponary a short distance away with numerous stingless beehives sourced from the local vicinity. Appended overleaf, an addendum of Perry Mateo's account.

7.6.5 Addendum - Mateo Meliponary; by Perry P. Mateo

Figure 109 Left: young Duco Lanzones fruits developing Right: volume & spacing of matured Longkong Lanzones fruits

Stingless bees or Kiwot as they are commonly called were not new to me, having been born and raised in Labo, Camarines Norte. They were always around on our small farm; however, we never entertained the idea to propagate them.

During my employment search, I was taught and given the chance to serve as an assistant beekeeper to Rico Pietro Omoyon who was then starting MILEA Bee Farm in San Jose, Batangas. The farm had been up and running for a couple of years when I came. I was also allowed to tag along as he gave beekeeping workshops around the country, and those were my first steps towards beekeeping exposure to both apiary and Meliponary. Rico gained his beekeeping know-how here and abroad and had a different perspective on his beekeeping approach. That would manifest even in his hive designs and now he is in the driver's seat of Omoyon Bee Farm in Dumaguete City, Negros Oriental. His beekeeping style inspired me to take on my flight for a stingless beekeeping journey here at Sabang, Pinamalayan, Oriental Mindoro.

Figure 110 Left: Perry P. Mateo inset inspecting ready-to-harvest Longkong Lanzones fruits; Right: Gavino G. Ermita standing beside a fruiting Dwarf Coconut tree

My father-in-law (Gavino G. Ermita) and my wife (Maureen) opted to start keeping bees here at Gavino Ermita Fruit Farms way back in 2013 with European bees that were brought seasonally for pollination in our farm locations. This happened after visiting Mr Omoyon in Batangas.

Fruit farms were in different locations and such farm operations on a large scale eventually extracted a toll on beekeeping, and at our location predators like the Bee Eaters also were a challenge.

I eventually took over the beekeeping side of the fruit farm last 2015, but this time focusing on stingless bees that were initially brought by the Mangyan tribesmen of Tau-buid. I propagated them following Rico Pietro Omoyon's hive designs, the 5 inches per side cube box. I had them lined up at the back portion of my in-law's residence while others were often transferred to our blooming farm locations. We were much amazed at the effects of pollination after being able to harvest more than half a ton of Duco Lanzones from just a single tree, all by using the stingless bees for pollination as shown in Figures 1 and 2.

Positive effects and the rise of crop yield due to stingless bee pollination were not only evident in our Lanzones trees but also other crops such as Coconuts (Figure 2), Rambutan, and Mangosteen trees.

This spiked our interest in focusing to increase the number of our colonies to four hundred (400) hives with Mr Omoyon visiting us from time to time for additional guidance.

Figure 111 The spillway adjacent to my area that locals oftentimes refer to as the Bridge Under Troubled Waters

We did not stop there, and before we knew it, we already had over two hundred fifty (250) hives in total. We had hives in Poblacion and Lagpan, Sabang, Pinamalayan, and Villaflor, San Teodoro all of which are in the Province of Oriental Mindoro. We also had them at my father-in-law's residence in early 2018. The typhoon came regularly, and we just shrug them off.

Figure 112 Left: hives with different measurements; Right: feral colony with coconut shell halves at right

We made sure that colonies are at their peak strength in terms of volume before they are relocated for pollination. We had everything prepared for both bees and fruit trees but failed to reinforce them for the typhoons. Little did we expect a sudden shift in typhoon strength. We forgot that Oriental Mindoro is a frequent typhoon path and that led to the downfall of our colony holdings to less than 200 colonies at one point.

Our colony numbers decreased drastically for every typhoon that came. Much as we would want to focus on the bees, we need to first address the devastation of our plants, which will eventually form part of the forage for our bees. Bringing back the glory of our colony holdings was no easy feat. It seemed like starting over again. We came to a point of giving up simply because we could not cope with the workload required to sustain beekeeping and pollination operations.

Whenever we got together to talk about the realities we were facing, it was always felt that we needed to heed Mr. Omoyon's advice that if we pursue fruit tree farming, we could not do it without the native pollinators. We decided to grow back our colony numbers but this time employing a different strategy, the 'budget-friendly' beekeeping way.

I was producing hives with cost savings in mind and that resulted in a hive box with no definite or standard measurement as shown in Figure 4. Hives succumbed to typhoon floodwaters and the strong winds but those also brought down some trees that we turned into logs and eventually lumber to restore damaged properties or to convert the same into cash. I then turned the scrap lumber into hives of different unique dimensions, as hive material is determined by the available lumber.

Whenever we encounter feral colonies in our farm sites, we relocate them to our secured beekeeping site to install in coconut shell halves and allow them to recover for the next three months before transferring the same to the box hive for migratory pollination. We have observed optimal harvest in pollen, honey, and propolis following this method.

Figure 113 multi-level hut for hives

Our hives are sheltered under one roof for ease of monitoring, especially during a typhoon. It is just a multi-level structure near coconut trees.

Currently, I have about 80 colonies surrounding my residence, just a few steps away from the spillway and the creek. Coconut trees abound as a forage source for the bees. We also have about 130 colonies at my in-law's residence, for the time being.

We intend to push our colony numbers forward from now on. We intend to bring back our glory days of stingless beekeeping, utilizing the learning we had from the past. We are now opening our farms to others who wish to learn from our expertise and experience in using stingless bees for pollinating our fruit trees.

It is now our mission to introduce stingless bees for pollination at least for the entire Mindoro Island using our farms as models for others to pursue and learn from, with or without government assistance.

7.7 Visiting Perry's father-in-law Gavino Ermita[8]

7.7.1 Getting there: The father-in-law's place is about 2 km farther in and we had to board his jeep because the path crosses a creek 7 times as there was no proper road.

There were approximately 100 hives with 80% in wooden box hives. All were obtained nearby. Notably, the aperture of most entrances seems much bigger than what we saw in Luzon. The hives that

[8] The group photo in front of his modified jeep to ferry passengers as well as his agricultural produce. Appended, a photo array of the nest entrances of beehives in Gavino's Meliponary.

we could examine had a strong healthy brood, a testament to the natural surroundings with no real development or pollution.

Figure 114 Top left: form left Neth Pastor, Perry and his children, Mar Bautista, Dr. David Roubik, Mrs & Mr Ermita, AHJ, Victor Samarita & NAP; Top right: Gavino's house; Bottom left: "road'/creek crossing leading to Gavino's place from Perry's place; Bottom right: one of Gavino's rack of SB hives

7.7.2 Photo array of Nest entrances at Gavino Ermita's Meliponary

Figure 115 There were approximately 100 odd hives with 80% in wooden box hives. All were obtained from nearby area

7.7.3 Healthy colonies[9]

Figure 116 The hives that we got to examine had strong healthy brood, a testament to the natural surroundings—no development and pollution

[9] All hives domesticated from feral colonies sourced in the local vicinity.

7.8 Visiting Victorino Samarita, Barangay Maragooc, Gloria

7.8.1 Getting there: About an hour's drive from Pinamalayan.

7.8.2 Etymology: Gloria means "glory", from the Portuguese and Spanish titles of the Virgin Mary Maria da Gloria and María de Gloria. We speculate that Gloria could be named after Gloria Macapagal, Maria Gloria Macaraeg Macapagal Arroyo, a Filipino academic and former politician who was the 10th Vice President of the Philippines and 14th President of the Philippines, from 2001 until 2010.

Figure 117 Victor has one hive of T. biroi that he brought from Luzon

Victor has one hive of *T. biroi* that he brought from Luzon when he was working there. Besides acting as the guide for this leg of the tour, he decided to bring us to his land, which he intends to convert into a Meliponary. Buildings are securely locked, and no one is living there when he is away. It is a decent piece of land that needs some clearing and a basic plan for starting a flower garden with a vegetable plot and possibly some local herbs as well. The main colony he has looks mature enough to be 'divided' so he can start building structures to accommodate racks and provide protection from direct sun and rain. He can gradually explore his area for any feral colonies that may exist there. After a while, we left and headed for Neth Pastor's Farm.

7.9 Ramona Mantaring Pastor, HN Organic Farm, Barangay Malitbog, Bongabong

7.9.1: Getting there: From Gloria, it is approximately a one-hour drive

7.9.2 Etymology: Malitbog - to pay homage to the Holy Child Jesus

Figure 118 Neth Pastor in front of her traditionally built house

7.9.3.1 Etymology: Bongabong - The town is home to Kuta Bongabong (Fuerza de Bongabong), one of the oldest Spanish colonial fortifications in the Philippines. The Ruins of the Church of Kuta or Fuerza de Bongabong are the remnants of an 18th-century church built by the Recollects in Bongabong, Oriental Mindoro.

7.9.3.2 Kuta Bongabong History

The Recollect mission of Bongabong was first mentioned in the Libro de Registros in 1737. The church served as a garrison for the townsfolk during Muslim raids. It was destroyed during the raids between 1753 and 1754. It is located at Sitio Cuta, Barangay Anilao, Bongabong, Oriental Mindoro. The Kuta ruins are across the Bongabong River mouth and about 26 min. away from HN Organic Farm in Malitbog.

7.9.4 HN Organic Farm History: I was curious and just had to ask how is it that her name as Ramona could turn out to be Neth. As it happens, she said that her name in her younger days was shortened to Mona and then to the likes of Monet and inevitably later even shorter to just Neth. This only tells me that this is a friendly and pleasant lady with many friends. Similarly, the farm was originally started by her parents and the name was shortened to simply the initials HN. The farm is about 19 hectares and the premise also serves as an agriculture training institute. It features herbals, horticulture, poultry produce and some forms of permaculture. With coconut trees all around though not projected as a coconut plantation, coconuts are among the farm Agricultural products. The stingless bees that she acquires from the region fit in well with this portion of agricultural produce.

Figure 119 HN Organic Farm and Agriculture Training Institute signage

7.9.4.1 Photos of HN[10] Organic Farm, Malitbog, Bongabong

Figure 120 The landscaping though sparse, does provide some welcome ambience for visitors and bees

[10] HN Organic farm is named after Honorio, the father and Nelcy, her mom's name; HN can also mean 'Harmony with nature'.

7.9.4.2 Stingless Bee colonies[11]

Figure 121 Assorted SB hives, some in coconut shells and hung, some in Bamboo, and some in wooden boxes

[11] See 7.9.5 Addendum Profile: Ramona Mantaring Pastor.

7.9.4.3 Termite-infested Branches[12]

Figure 122 Top row: Coconut shell kernel hives; middle and bottom are termite nests in fallen branches, with stingless bee nests

[12] According to Neth, all her SB colonies were sourced from the local vicinity.

7.9.4.4 Assorted Beehives

Figure 123 Appears to me Ms. Neth collects first all manner of hives she can acquire and then arranges them according to type and material and finish. Some sections will house coco shell kernel hives, some with white painted wooden box hives and some plain lumber

Figure 124 Stingless bee hives are spread all over the 19-hectare Farm

7.9.5 Addendum - Profile: Ramona Mantaring Pastor

Neth Pastor is a farm operator from the centre of organic farming in Oriental Mindoro, Bongabong, Philippines. She was formerly a public servant at Overseas Worker Welfare Administration before joining her family abroad. When she returned to the Philippines, because of her love for farming, she developed the HN Organic farm. The farm is situated at km 5 from the town proper. It is a working farm dedicated to organic farming activities. It is accredited by the Agricultural Training Institute of the Department of Agriculture as one of its Learning Sites for Agriculture partnering with their e-extension activities catering to farmers, community leaders, and agricultural students alike. HN organic farm's original objective was to maintain the natural serenity of the place for the family's retreat and vacations. It includes the propagation of native stingless bees. Ramona Pastor got interested in beekeeping during her younger years because of her grandparents who had full respect for the environment. His parents who started to develop the farm during their free time from their busy teaching jobs supported her when she returned from employment abroad.

She took intensive beekeeping training from the University of the Philippines when she started managing the farm. She secured a National certification for Agriculture Production qualifying her as Trainer and organic farm practitioner.

Their bees are preserved in the secluded areas where they were originally found them. You can see the bees in their different habitats around the farm. She wants the farmers to appreciate the bees and explain to them how they can help our ecosystem. She wants to educate farmers on how they can integrate the bees to increase their productivity by pollinating their crops.

With her knowledge, training and experience as a beekeeper, she can share with the farmers why bees are important and how agronomist people can lend a hand to our worldwide concern and continuing campaign to Help Save the bees. The farm is open for educational tours. It is a unique educational experience that both kids and adults enjoy. They can have a taste of honey fresh from the beehives and know the sources of nectar and pollen as well as edible flowers. They also discover the relationship between bees and flowers and how bees are closely tied to the food chain. She can also share her knowledge of how we can derive other bee products aside from honey, the last expensive form of bee products. There are a lot more high-value products we can produce from bee propolis and wax. The farm is very much conducive to a learning experience about native bees.

She firmly believed that with the Agricultural Training Institute's (ATI) integration the farm will have a big impact on her campaign in preserving the native bee species on the island of Mindoro.

7.9.5.1 Profile: HN Organic Farm grows and produces in harmony with nature. Here we make different natural concoctions from the farm's raw materials to produce organic fertilizers. We focus on soil fertility. What we put in is what we get back. We care for livestock in their natural way of living. We use natural pests and insect repellents for destructive insects and care for useful and helpful ones like bees. Most importantly we use stingless bees as our great pollinators to increase our farm production. We propagate them to take advantage of their high-value products. We obtain from them pure and raw honey, bee pollen, beeswax, bee brood and propolis. These are very much in demand globally.

Climate change is the biggest hindrance to beekeeping, and we should be serious about its impact, like the COVID-19 pandemic. At HN every aspect of life should not be taken for granted.

Chapter 8 - Occidental Mindoro

8.1 Occidental Mindoro State College (OMSC), San Jose City

8.1.1 Road Trip: Victor was genuinely concerned about the road conditions to get to San Jose City from Hagan where we put up the night. Mar Bautista our driver knew the road and said that it would take 2.5 hr. drive time in normal conditions. We left at 4 a.m. and expected to get there before 7 a.m. Donato Cuden would be waiting for us in San Jose City Square. Approaching Magsaysay, we saw the concern, it was a steep decline through the winding road from the highland, although the view was exhilarating. Good thing the weather was favourable at that time just after sunrise. We slept in the vehicle most of the way whilst Mar was alert enough all the way. We stopped for a breather just before descending the steep road. We took some nice photos of the view of Magsaysay town from the peak of the road decline. The descent was quite daunting, Mar took it slow and steady, and we arrived safely.

8.1.2 Etymology: San Jose - From Spanish San José ("Saint Joseph") (Anglicized with the removal of accent)

Figure 125 Just before descending the steep road, NAP took a photo of the view. Magsaysay town in the background.

We arrived at San Jose City square way before 7 a.m. and no one was there yet. We decided to browse the area while waiting for Donato Cuden who was going to guide us throughout Occidental Mindoro. He arrived shortly with Zeny Topado, Rose Cuden and Alrizza Zubiri. From the square, we headed to the nearest MacDonald's for some long-awaited breakfast.

We were supposed to meet with Dr. Gilmar of OMSC but Donato said that regrettably, he was not available. We decided to fill the time gap by visiting Ms. Zeny although she was initially reluctant to bring us to her place because it was recently devastated by the typhoon.

8.1.3 Getting there: After breakfast, we travelled for about 15 minutes in 2 vehicles with Donato and the OMNC staff guided the way to the agriculture school Meliponary. There we were introduced to the custodian, Dr. Arnold Peralta.

8.2 Visiting Dr. Arnold de Peralta of OMNC Agricultural School, Brgy. Canwali, San Jose, Meliponary

Looking at many clusters' entrances from a different region, there seemed to be a peculiar pattern in general or mode of formation, particularly for this *T. biroi*, if they indeed are a single species, especially considering those in Oriental Mindoro.

Figure 126 Met up with Dr. Arnold Peralta, the custodian of the OMNC Agricultural school Meliponary

8.3 Visiting Zeny Topado, Bubog, San Jose

Figure 127 Some box hive colonies on the racks of the Meliponary shack. Group photo: there were 2 cameras, with some looking at one and others looking in another direction. Anyway, from left, Ben, Arnold Peralta, DWR, Zeny Topado, NAP, Donato Cuden, AHJ, Dr. Gary Calitang, Mar Bautista, Rose Cuden, Alrizza Zubiri, Neth Pastor and Victorino Samarita in front.

Figure 128 We arrived at Ms. Zeny's place to observe almost total destruction by typhoon

8.3.1 Getting there: It was a short ride to Barangay Bubog, a village on the outskirts of San Jose

8.3.2 Etymology: Meaning of *bubog* in Tagalog. = n. 1. crystal; 2. broken glass

8.3.3 Typhoon devastation - We arrived at Ms. Zeny's place to observe almost total destruction[13] from the recent typhoon. Even in that adversity she still wanted to serve us coconuts as refreshments. Filipino hospitality is at its peak.

I guess in normal circumstances such hive failures would have been quickly forgotten. However, with her difficulty in sourcing hives and her time constraints, she was probably trying to salvage the wooden boxes. At this juncture, I would suggest removing everything and refinishing the inside with ethanol extract of propolis, setting it to dry thoroughly and then flame the inside surfaces with a torch.

For future precautions, Ms. Zeny will need to securely anchor her hives and provide better structural protection from the high winds and heavy rain. The garden itself could do with some screen planting at the perimeter.

8.3.4 Dead colonies in Typhoon aftermath

Figure 129 Typhoon aftermath left the box hives infested with fungus. Further retardation by cockroaches

[13] Appended overleaf, photos of the insides of Ms. Zeny's box hives. Fig. 57 Dead colonies in Typhoon aftermath.

8.5 Donato S. Cuden[14] and Rose S. Cuden 1750

Teachers Village, Brgy. Lubanga Población, San Jose

We did not get to visit Donato's place, but he contributed a photo of the hive he owned. It is a box hive of *T. biroi*. He has a bowl of water with a rock in it to quench his bees' thirst, I guess.

Figure 130 Donato's Hive

[14] Donato S. Cuden was instrumental to our whole itinerary in Occidental Mindoro.

8.6 Visiting Danny Viola Hizon, Danizon Farm

Sitio Inagbaan, Brgy Poblacion, Sablayan, Occidental Mindoro

8.6.1 Getting there: A one and half-hour drive northwards from Bubog along the western coast of Mindoro

8.6.2 Etymology: **Sablayan** was derived from the word sablay, a Visayan term **meaning** wave convergence. In the early times, the very location of the town was where the waves from the North and South China Sea meet, hence, the name Sablay which later became Sablayan.

Figure 131 Drinking in the beautiful scenery along the way we soon arrived and were greeted by Danny Hizon, owner of Danizon Farms

Drinking in the beautiful scenery along the way we soon arrived and were greeted by Danny Hizon, owner of Danizon Farms. The farm boasts 16 hectares of organic agriculture. The first thing Danny did was invite us to his meeting pavilion where he had some lunch ready for us. It was a hot day and we gladly quenched our thirst with some cool drinks.

His house was where he had all his stingless colonies. He has been practising colony multiplication due to scarcity and difficulty to source additional colonies.

Figure 132 The first thing Danny did was invite us to his meeting pavilion where he had some lunch ready for us. It was a hot day and we gladly quenched our thirst with some cool drinks

Figure 133 Danny's house up an adjacent hill. The structures are traditionally architecture and incorporated bamboo and attap roofs.

8.6.2.1 Danizon Farms Photo Gallery

Figure 134 After a short tour of his farmland, we were invited to his house up an adjacent hill

8.6.2.2 Horticulture and Meliponiculture integration[15]

Figure 135 Horticulture and Meliponiculture integration

[15] see 8.6.3 Addendum Danizon Farms.

8.6.2.3 Benefits of Semi Aquatic plants

Figure 136 Semi-aquatic plants with tiny flowers, and water, provide for SB foraging

8.6.3 Addendum - Danizon Farms by Danny Viola Hizon

Danizon Farms, located in Sablayan, Occidental Mindoro, is an integrated, diversified organic farm that was established on July 1, 2015. It was accredited by the Agricultural Training Institute in April 2018 as a Learning Site for Agriculture. Currently, it is working to complete the requirements by the Department of Tourism for accreditation as a farm tourism destination.

The farm sits on a 16-hectare agro-forest land. Gradual clearing of the land is being done to accommodate the planting of diverse fruit-bearing trees. As of this writing,

Image 15 Bird's eye view of the farm

the plantings include embryo-cultured macapuno coconut, Malaysian latex-less jackfruit, pomelo (Davao "Magallanes", Malaysian and Thai varieties), Rongrein rambutan, Bangkok sweet tamarind, dragon fruit, guava (Vietnamese pink and Thai guapple varieties), Sweet Elena mango, cacao, coffee, calamansi, cashew, papaya, banana, and specimens of other fruit-bearing trees and vines. In addition, small areas are devoted to organic rice, vegetables, root crops, herbs and spices. Livestock includes free-range chickens and cattle.

The facilities existing on the farm are a farmhouse, a multi-purpose hall, two native cottages for guest accommodation, a poultry house, a cattle shed, a vermiculture shed, a small Meliponary, and a small hall with ocean and mountain views, a concrete water tank, and a nursery. In the works are two greenhouses for vegetable production, a fishpond, and a natural swimming pool. In the near future, more guest cottages and a larger vermiculture facility will be built, a 5-hectare forest will be upgraded by planting native hardwood trees, a one-hectare bamboo garden and a butterfly garden will be established, and a campsite and forest trail will be installed.

Image 16 Farm scenes – 16 hectares of farm land open for educational forest trails

The farm is the realization of the owner's dream of a tranquil retirement place in a mountain setting. It is intended to achieve two personal goals of his. Firstly, to honour his farmer parents and especially his grandfather who tried to make him a farmer during his teen years but failed because the teen's interest was in schooling. This farm is to say to him “I am sorry I disappointed you then but now looking down on me from heaven I hope I make you proud”. And secondly, to show all that it is never too late to start farming.

Image 17 Farm products – Agriculture produce; Papayas, Coconuts, Dragon fruits, Pomelos and ruminants.

Image 18 More farm products like grapes and pomelo plus processed goods

8.7 Visiting Lumangbayan Road, Santacruz

Figure 137 We arrived at Lumangbayan Road, SantaCruz and stopped by the house of Fernando Tuzcano. Bees were foraging cashew nut flowers outside his fence.

8.7.1 Getting there: After about one hour's drive from Sablayan, we arrived at Lumangbayan Road, Santa Cruz and stopped by the house of Fernando Tuzcano, the chair of Occ. Mindoro Beekeepers Association

8.7.2 Etymology: Lumangbayan Road means the road leading to Lumang Bayan in the city of Calapan, province of Oriental Mindoro. Lumangbayan, meaning "former población". This is the junction to the way to Barangay Santa Cruz (or **Santacruz**) and a Spanish or Portuguese term **meaning** "holy cross" and refers to the cross on which Jesus was crucified

8.7.3 Residence[16]: He was not in and we could only view his hives outside. There were a few cashew nut trees that were blossoming and fruiting. We only were able to photograph his garden from outside the fence.

8.8 Visiting JOPSON FARM Brgy. 8, Mamburao, Occidental Mindoro.

8.8.1 Getting there: From Lumangbayan Road we headed north for another hour or so and arrived at Jopson Farm in Mamburao.

8.8.2 Etymology: Mamburao was originally settled by the Moro people and was one of their strongholds on the island of Mindoro. The settlement was the first one in Mindoro to be plundered by the Spaniards.

8.8.3 Jopson Farm: Dr. Nathaniel Descanzo[17] was not available but we were met by Dr. Adrian de Ramos. One of the racks had a bowl of water and some rocks in it to make sure the bees do not drown.

8.8.4 Occidental Mindoro Beekeepers Association (OMBA)

Figure 138 Left: Dr. Adrian and Donato showing AHJ the colonies. Right: a bowl of water and some rocks in it to make sure the bees do not drown.

[16] The Fernando Tuzcano residence is also the Occidental Mindoro beekeepers Association and he is the chairman of the association.

[17] Dr. Nathaniel Descanzo was introduced to us by Dr. P.J. de Castro of Manuel S. Enverga University Foundation, Lucena City, Philippines.

Figure 139 Top left: Bee Hives from the Occ. Min. Beekeepers Association. Top right: Viewing the racks; Bottom left: one of the healthy T. biroi colonies; Bottom right DWR checking out the map and Posters

It is already dark, and we were in a hurry not to miss our bus ride back to Manila; we didn't stop for a breather. That picture was taken after we visit the OM Beekeepers Assoc. President's house and we dropped by because Ms. Alrizza wanted to show us the mountain where natives live. Then we moved towards Paluan in NE Mindoro.

Figure 140 Jopson Farm and Meliponary

8.9 Manyang Iraya Community @Ulasan, Brgy Harrison, Paluan

8.9.2 Paluan Indigenous people:

Natives Tribe Iraya of Calawagan Mountain, Paluan, Occidental Mindoro

We arrived at Paluan around 5:30 p.m. with the help of Fely Egos, the Municipal Agricultural Officer of Paluan, and I had a chance to interview two native bee hunters, namely Clifford Parisan and Solomon Sanbutan. According to them, they are harvesting *Apis breviligula* honey and they also have seen the stingless bees that they call "Mahsam".

After the interview, we agreed that whenever they could get stingless bee colonies, they will keep them and report to the Municipal Agriculturist so that they will be given formal training in keeping them.

It was almost 6 p.m. when we left. We stopped for a while at Calawigan Mountain Resort. It was beautiful and suited for family bonding, especially during weekends and summer vacations.

Figure 141 Apis breviligula Philippine giant honeybee by DWR

Figure 142 At Mamburao we stopped at the coast to view the location of Paluan i.e., the North-eastern tip of Mindoro

The indigenous people of Oriental Mindoro are the Mangyans (Manguianes in Spanish, *Mañguiane* in Old Tagalog), consisting of seven distinct tribes. They occupy the interior, especially the highlands. Mangyans have inhabited the island since pre-history. They are believed to have originally travelled from Indonesia and settled down on the island. Mangyans were the ancient aborigines of Mindoro.

Even now, the Mangyans that are of Malayan descent are known to be the first inhabitants and were later pushed to the interior mountain areas due to the arrival of Christian settlers.

The tribes, which are referred to by the general term "Mangyan", comprise the Alangan, Bangon, Buhid, Hanunoo, Iraya, Ratagnon, Tadyawan, and Tawbuid. The Mangyan population in the Philippines is over 100,000, and the great majority have put down roots in Mindoro.

The Iraya Mangyans live in Occidental Mindoro, mainly in the towns of Abra de Ilog, Mamburao, and Paluan. Some populations of Iraya are also found in the Oriental Mindoro towns of Puerto Galera, San Teodoro, and Baco.

The Iraya is usually not concentrated in just one specific area; they live in various barangays* of a municipality. Their population in Puerto Galera, northwest Mindoro, is a typical example. Barangay Baclayán, which is about a one-hour walk from Santo Niño, has around 600 Iraya. The barangay has a health clinic and a school, and a very active community.

Traditionally, the Mangyan are animists and are known to be superstitious. Today, however, many Iraya are Roman Catholic, Evangelical Christian, and Iglesia ni Cristo, which is attracting a growing number of converts from other faiths.

8.9.3 Ulasan, Brgy Harrison, Paluan

8.9.4 Etymology: Ulasan is the name of the community. Is that a Mangyan Iraya word? In Malay "ulasan" means "expression". I suspect their language may have many Malay words. In any case, palu in Malay means to hit with a stick... Paluan = result of hitting with a thick rod or pole.

Figure 143 From left, NAP, seated Alrizza Zubiri, Clifford Parisan and Solomon Sanbutan, Donato, AHJ; Standing from left: Mar Bautista, Rose Cuden, Zeny Topado, Neth Pastor, Fely Egos, Victor, DWR, Canadian visitor.

8.9.5 Getting there: Anyway, this destination was not originally on the itinerary. The idea to visit the town Paluan and subsequently the indigenous community was mooted when we were inquiring about endemic species of bees in Mindoro. This destination, at the north-eastern tip of Mindoro, we were told, may prove impossible because of the distance and time constraints that we faced. It would take about 3 hr. to go and return to Mamburao.

Incidentally, the Mamburao Bus terminal is where we were to end our Mindoro tour. We were to alight the RORO bus here to proceed to Manila for our next leg, to Palawan.

8.9.6 The Ulasan community - With much insistence from us, Ms. Alrizza Zubiri made a few phone calls and arranged for us to meet with the Ulusan community elders for an interview with their bee hunters. We made our way to the Paluan Municipality and met up with Ms. Fely Egos, the Municipal Agricultural Officer. She accompanied us to the village and introduced us to the community elders.

We all gathered at the Church front courtyard and among the honey hunters, only two of whom appeared. We were grateful that we got to meet them because it was an impromptu visit. Ms. Alrizza introduced us to Ms. Fely and conveyed our quest to gather information about the bees and honey that they gather. As a gesture of goodwill, I suggested to NAP that we could buy some honey if they had any. This would at least provide some gain for them from our visit there.

Anyway, Ms. Fely explained our purpose there to the matriarch of the community and requested to meet and interview the honey gatherers or bee hunters. Ms. Alrizza proceeded to talk to them, and NAP interviewed them regarding their activities and most importantly whether they chanced upon any stingless bees.

As expected, they mentioned that they do encounter the SBs and, in their language, they called the stingless bees “Mahsam”. The name was not a big surprise because the word Masam in Malay means sour. The taste of SB honey would be sour as compared to honeybee honey.

8.9.7 Barangay Harrison, Paluan[18]

Figure 144 Top photos: signages of the Trading post and training centre at the Paluan Municipality Agriculture office. Middle row: Interesting feature trees in the village give a cosy landscape. Bottom photo: The Church front courtyard where we met with part of the community.

[18] The agriculture activities are monitored by the Paluan Municipality Agriculture Office.

8.9.8 Encounter the Iraya Manyang Tribe[19]

Figure 145 Ms Alrizza, Ms Fely and NAP discussing with the honey hunters accompanied by the Matriarch of the community.

[19] We were greeted by the matriarch of the community.

Chapter 9 – Palawan Island

Map 2 Routes of Puerto Princessa, Palawan Tour Mar. 2020

9.0 Palawan Indigenous people**:** The native-born Palaweños still dominate the populace. Eighteen percent is composed of cultural minority groups such as the Tagbanwa, Palawano, Batak, and Molbog. Incidentally, there is a Batak community among the native peoples of Sumatra.

9.1 Lelet Tabinga Stoltz @ Haley's Farm Gabinete Road, Bancao-bancao, PPS

We arrive at Puerto Princesa (PPS) airport and were met by Joseph Dador (a friend of a friend) with Jun Tabinga (the brother of Ms. Lelet, the owner of Haley's Farm)

9.1.1 Getting there: An acquaintance Joseph Dador (nicknamed "Jok") brought a vehicle that he said we could hire and for a fee, he will drive us around for 2 days. We all agreed and proceeded to Haley's farm which was about 10 or 15 min. away from the Airport.

9.1.2 Etymology: Puerto Princessa – Historically, the place was named after Maria Cristina, one of the princesses born to Isabella II of Spain and her consort, Francis, Duke of Cádiz. When the princess suffered an untimely death, the Queen changed the name to Puerto de la Princesa. Eventually, the name was shortened to Puerto Princesa.

9.1.3 Haley's Farm: Ms Lelet's beekeeping farm was named after her daughter Haley. It constitutes a garden with many floral blooms unique to Palawan.

As expected, the variety of plants and trees in Palawan would be a tad different because of its geographical location within the Wallace line. Palawan being on the west of that line would have flora and fauna different from the rest of the Philippines, more like Peninsular Malaysia or Borneo. By the same token, one would expect the diversity of bees would also differ to some extent.

There is one stingless bee log hive, wrongly cut too short at the bottom, and we could see the inside by just tilting the log. Examining the brood we suspect that it could be *Tetragonula* cf. *sapiens*. *Apis mellifera* colonies in wooden box hives are at the back and the front office and there are racks of honey products, and a show tricycle painted with a beekeeping scene.

Figure 146 Different vegetation that is not normally seen in other Philippine islands

In the meantime, Dr. Roubik had spotted an *Apis andreniformis*[20] colony in the garden which came as a surprise to Ms. Lelet because she was unaware of it being there. Done with our visit, we bid farewell to Ms. Lelet and her daughter. Then, with Ms. Lelet's brother Jun who is now our driver and guide, we left Haley's Farm and headed northwards towards Sabang Beach.

[20] In the other Philippine Islands, it is quite common to see *Apis breviligula (see* Fig 85, *Philippine giant honey bee) but here in Palawan in plain sight is an Apis andreniformis colony. (See photos overleaf).*

9.1.4 Palawan Blooms[21]

Figure 147 Some of the floral blooms unique to Palawan Island

9.1.5 Palawan Stingless Bees

Figure 148 Inside a T. cf. sapiens log hive looking in from the bottom

Figure 149 Top left: Apis mellifera colonies in wooden box hives Top right: a view from inside the office; Mid left Racks of Honey products; Mid right: A show tricycle painted with beekeeping scene; Bottom left an Apis andreniformis colony spotted by DWR in the garden. Bottom left; Jun Tabinga, NAP, AHJ, DWR & Lelet Stoltz

9.1.6 Addendum - Haley's Bee Farm; by Enrelita "Lelet" Tabinga-Stolz

Gabinete Road, Puerto Princesa City, Palawan

My beekeeping journey started back in 2005 when I attended a training program by the then Mayor of Puerto Princesa City, Edward S. Hagedorn. I was able to meet experienced local beekeepers and enthusiastic members who were keen to share their knowledge. The training aroused my interest in beekeeping and inspired me to pursue it. I never knew that this, which I once thought will just be a hobby will turn out to be a passion of mine.

Image 19 My apiary focuses on Apis mellifera

After I completed the training and with my good friend Maria Fiechter's encouragement, I decided to give beekeeping a try. I purchased my first hive and started in the backyard of our residence located in Barangay Bancao-bancao, Puerto Princesa City, Palawan. I named it after my youngest daughter, Haley, who is growing up exposed to beekeeping. My first year was not easy. I struggled to apply my newly acquired theory in practice, at the same time trying to keep my hive alive. At one point, I even lost my queen, and this made me almost give up beekeeping. Our mentor, Christine Dador, persuaded and helped me to keep ongoing.

In 2010 we faced a drawback which affected a lot of apiaries in the city. Due to a pest called small hive beetle *(Aethina tumida),* we were not able to get fertilized queen from Manila or abroad. This opened an opportunity for me to learn how to produce my queen, which helped sustain my 10 colonies. I continued producing my queens for 3 years until I was able to get a new imported queen. And the beekeeping community started to flourish again. Interest in training grew.

Before I started beekeeping, I enjoyed working in the garden. It was my hobby back then. I have always loved flowers and plants. Little did I know that this would eventually help my apiary to flourish. The abundance of different floral blooms, mostly floral species that are unique to Palawan, in our garden has attracted a diversity of bees. My apiary focuses on *Apis mellifera*, however, other bees such as

Image 21 my 10-year-old daughter, Haley, has been helping me around during honey extraction

Fifteen years later and after a fair share of challenges and a lot of hard work, what began as a hobby has now grown into a business with approximately 21 hives in production. I was able to slowly expand my colonies, still in our backyard. During each honey flow season, I'm able to extract around 300 to 400 kilos of honey. My family, especially my 10-year-old daughter, Haley, has been helping me during honey extraction. I am also able to support the indigenous people by purchasing their wild honey (*Apis breviligula*) since 2009. Apart from managing my apiary, I am conducting training and selling colonies to those who would like to pursue beekeeping. I offer educational tours to schools and help educate the students about the importance of bees in our ecosystem.

Image 20 I offer educational tours for schools, to help educate the students about the importance of bees in our ecosystem

Beekeeping has opened a lot of opportunities for me. I meet a lot of people who are keen to share their experience and knowledge, from European beekeeping to tropical beekeeping. It has become a very enjoyable and sociable pastime, visiting others' hives, picking up tips and hearing success stories, not to mention having a sustainable livelihood.

9.2 Visiting Yamang Bukid Farm, Sitio Candia 3, Brgy. Bacungan, Puerto Princesa

9.2.1 Getting there: About 40 min. into our journey, we arrived at Brgy Bacungan and decided to visit Yamang Bukid Farm. There has been much hype about this Agro-Tourism Farm so we thought we would see what it's all about. The farm is about 10 min. from the main road along the Bacungan River.

9.2.2 Etymology: Bacungan means "barring" in English. Possibly from the fact that the Barangay is barred by the Bacungan river and a lush Mangrove forest along with it and surrounding hills.

9.2.3 Etymology: Yamang Bukid, according to the online Tagalog dictionary (https://tagalog.pinoydictionary.com/), **yamang** = conj. 1. since; because; 2. whereas; 3. now; **bukid** = n. 1. rural areas; 2. farm; field under cultivation

9.2.4 Yamang Bukid Farm - The Yamang Bukid farm in Puerto Princesa City, Palawan was intended to be a turmeric plantation, but due to the evolving needs of the local farmers and the community of Barangay Bacungan, it has turned into a place for relaxation and habitat to wild, endemic species of animals and insects in the area. The latest reports indicate Yamang Bukid is a 10-hectare farm.

9.2.5 Meliponiculture**:** The spread was overwhelming. We had to walk around a few different sections of the farm and inquire from several staff and workers until we came across the Meliponary. The feature structures were unique and quite impressive but the Meliponary had much to be improved. It is clear to us that the farm management is either not too keen or does not have much knowledge of stingless beekeeping[22].

The boxes were not well arranged on the rack, with empty boxes which appeared to contain failed colonies, and those still surviving. They are surviving where one would expect them to be thriving. We conclude that they were not well managed.

It was soon drizzling and we had to dash off for cover, with no time to examine the hives further.

[22] The Stingless bee Production Project started in 2019 and made possible through the UPLB Bee Program. Initial startup of 12 colonies and currently the 54 colonies came through acquisitions from the said Bee Program. The management establishes facilities and equipment to attain sustainable production of high-grade stingless bee honey. (Rodolfo O. Abalus, Jr., Email: rodolfo_abalus@yahoo.com).

9.2.6 Agro Tourism

Figure 150 Features and structures

Figure 151 A 10 hectares farm spread with a Meliponary station of T. biroi hives brought over from Luzon

9.3 Keith O. Mikkelson, Aloha Ranch and Organic Farm,

Brgy Macarascas, Puerto Princesa City, Palawan.

9.3.1 Getting there: About halfway through our destination of Bambua Nature Cottages at Sabang Beach in Cabayugan we arrived at Macasrascas.

9.3.2 Macarascas is a barangay in the city of Puerto Princesa, MIMAROPA, Philippines... coordinates are estimated at 11.2 meters or 36.7 feet above mean sea level. Population (2015): 1,609.

9.3.3 Aloha Ranch and Organic Farm

We stopped by this farm along the way when we saw the sign just at the side of the main road. The gate was closed but we explained to the caretaker that we intended to merely view the farm and possibly check out their beehives. It appeared the farm was also into hydroponics with assorted crops and plants. The bees were at the back and appeared to be local stingless bees.

Figure 152 The signage that caught our eye

The hives though only 3 in number were enough to indicate something of endemic Palawan Stingless Bees. At first glance, the nest entrance with its radiating ridges around the entrance had an aperture that reminded me of *Tetragonula testaceitarsis* found in Borneo and Peninsular Malaysia. We were eager to view the internal nest structure but unfortunately, the owner was not around so we had to content ourselves with photographs of the bees from outside the hive. One of the hives had a gap at its cover and the bees were actively going in and out of that gap. The boxes had clear plastic for observation under the lids but were almost fully covered with propolis so we were not able to view the internal nest structure.

Having taken our fill of photos, we thanked the caretaker and proceeded to our next destination.

Figure 155 They have an open godown and hydroponics set up

Figure 153 The Hive boxes; only 3 were enough to indicate to us an endemic Palawan Stingless bee. At first glance, the nest entrance with its radiating ridges around the entrance aperture reminded me of the Tetragonula testaceitarsis found in Borneo and Peninsular Malaysia.

Figure 154 Stingless bee colony photos from https://www.slideshare.net/KeithMikkelson2/apis-trigona-biroi-palawan?fbclid=IwAR1HE5y7eEvb_c1YqR8ehO9CPEBDviWYBo4-w5uj51USa6FUUtq7B0Z8WQQ

9.4 Iky Montebon Bacong, Brgy. Cabayugan, Sabang, Palawan

9.4.1 Getting there: By the time we arrived at Cabayugan, we all were hungry as it was lunchtime. Sabang is a small village or sitio located on the western coast of the provincial island of Palawan, Philippines. It is within the city of Puerto Princesa belonging to Barangay Cabayugan Sabang and is commonly visited by tourists as an access point to Puerto Princesa Subterranean River National Park.

9.4.2 Etymology: Sabang - The word “Sabang” refers to a place that can be found at the mouth of a river. Sabang is located at the very mouth of this subterranean river.

9.4.3 Beachfront: There were a few restaurants to choose from, but we were not too keen on the local dishes as most of the specialities were not Kosher or Muslim friendly. Jun brought us to Asiano beach restaurant that serves seafood cuisine. We placed our orders and while waiting, I decided to explore the village and vegetation behind the restaurant.

Thereupon I saw trees near the traditionally built bamboo houses, with characteristics that stingless bees would favour. I decided to call NAP to help me communicate with the residents about what we were looking for. True enough as it turns out, this lady, Iky Montebon Bacong has found bees by her house. We found a feral SB nest entrance at the crevice of an *Albizia falcata* tree; closeups of look like *Tetragonula*.

NAP explained the benefits of stingless bees to Ms Iky and provided an idea of how to possibly educt a colony into a hive box. We went back to the Asiano eatery to join DWR for lunch.

Figure 156 The seafront at Asiano eatery, bottom right: the village behind the restaurant

9.4.4 Barangay Cabayugan

Figure 157 Top row: DWR at the upper deck of the restaurant and the view from that upper deck. Mid row: from left NAP, AHJ and Iky Montebon Bacong; right: Iky's house behind Asiano restaurant; Bottom row; a feral SB nest entrance at the crevice of an Albizia falcata tree with some closeups of what look to be Tetragonula sp.

9.5 Kevin Dartman, Bambua Nature Cottages, Cabayugan, Sabang, Palawan

9.5.1 Getting there: From the Beach eatery, it was just a 5 min. drive to our Cottage gateway. However, it was a good walk up an incline to the actual cottages.

9.5.2 Etymology: Cabayugan means a place where bayug (a kind of bamboo) grows, hence the name.

9.5.3 Bambua Nature Cottages

Figure 158 Nest entrances in the joints between the curb and flooring finish in front of rooms

That evening we settled in at Bambua Nature Cottages and I was intending to explore the surrounding vegetation but what we found was the area was beside a rice field. Having not found any SB colonies and retiring, from the fact that I will not find any because of the rice fields, I returned to the room and had a chat with the owner's son, Kevin. Kevin showed us a honeybee colony in a palm tree nearby. With that, it set my hopes in anew, then I suddenly saw some bees flying with corbiculate legs hanging down, typical of a hovering SB. Looking around more intently, I found a few entrances at the joints between the corridor curb and the flooring pebble finish in front of the rooms. The bees appeared to be *Tetragonula* sp.

I was too exhausted by that evening and retired for the day after dinner. The next day after settling our accommodation bills at Bambua Nature Cottages we set out to look for breakfast along the way to explore south of Puerto Princesa.

9.6 Sammy Dagas, Kasapi ng Katutubong Tagbanua, Brgy. Buenavista, Puerto Princesa, Palawan

9.6.1 Getting there: Buenavista is a lookout point along the way about 25 minutes from Sabang.

9.6.2 Etymology: Buenavista means ”good view” in Spanish and in this context the location gives an excellent view overlooking the sea.

Figure 159 The Magayen IT Buenavista Consumers Coop; Bottom left: group photo, from left NAP, Jun Tabinga, Sammy Dagas with his child. Right: AHJ & Jun having a ciggy break at the Brgy Salvacion Signage

9.6.3 Tagbanwa Native[23] Barangay

The Magayen IT Buenavista Consumers Coop runs a souvenir and gift shop that I presume includes handicraft products from the Tagbanwa community.

We bought some souvenirs and handicrafts while DWR had some breakfast. The menu did not appeal to the rest of us so we continued to Barangay Salvacion nearby where I could savour some vegetarian breakfast. While at Buenavista, Jun introduced us to Sammy Dagas, a member of the Tagbanwa native

[23] See 9.0 Palawan Indigenous people.

tribe. Tagbanwa is a member of one of the oldest ethnic groups in the Philippines, mainly found in central and northern Palawan.

Sammy was confident he knew about the stingless bees we were talking about and he was positively encouraging us to follow him into his village and explore the vegetation there. Time constraints did not allow us to follow this suggestion. We need to plan another expedition for this purpose.

9.7 Perfecto Dias, DosfourDos Integrated Farm, Brgy. Magsaysay, Aborlan, Palawan

9.7.1 Getting there: To get to The Aborlan Municipal Hall **it takes about 2 hr. from Buenavista.** Arriving there we inquired at a grocery store for the Brgy Magsaysay.

9.7.2 Etymology: Aborlan – in Spanish means they tackle or overcome. Aborlan, officially the Municipality of Aborlan, (Tagalog: Bayan ng Aborlan), is a 1st class municipality in the Province of Palawan, Philippines. According to the 2015 census, it has a population of 35,091 people.

The barangay Ramon Magsaysay is named after Ramón [rah-mohn; Spanish rah-mawn], 1907–57, Philippine statesman: President Ramon Magsaysay 1953–57.

9.7.3 Profile: DosfourDos Integrated Farm, Brgy. Magsaysay, Aborlan. by Perfecto Diaz

Dospordos in Spanish means two by two, just some Spanish trivia, though the Owners may have a different reason for naming DosfourDos. We can only speculate but this is what the owner calls "2 sons and 2 daughters"... DOS. What to do, for example, The DOS of planting Dragon fruit, The DOS of Planting Cacao, etc. we will release a simple techno guide on the DOS. FOUR. God-given, nature-given..L-Light, A-Air, W-Water, S-Sunlight.

LOGO: 2 hearts: represents me and wife, (husband and wife) also, animals, poultry, ruminants and other wildlife on the farm that take breaths. Four leaves: represent 4 children, also, plants, life, and other living wild plants, on the farm that live or grow. Logo overall meaning: LOVE OF Farming.

We arrived at the farm with directions NAP got from Perfecto Dias, the owner. While waiting for the owners we explored the area and saw some nest entrances in a roof beam and some logs at the side of one of the structures. Managed to get a closeup of the bees at the roof beam and it appeared to be *Tetragonula iridipennis.* Dr. Roubik tried to examine some bees at the logs at the side while I managed to get one of the workers to cut the PVC pipe lengthwise, NAP told them that I will help them transfer the nest into a bamboo piece.

Figure 161 The DosFourdos emblem of two hearts with four leaves.

Having done that (i.e. splitting the SB colony in a PVC pipe and transferring it into bamboo), the worker said there was another nest at the top of the water tank tower. The water tank tower is about 12 meters from the ground and the rafters are another 3 meters. Looking at the struts, it had a fixed ladder at one of the tower pillars. I tested the strength it and it appeared quite stable I then decided it would easily take my weight. I ended up climbing the water tower to examine SB nest entrances in the rafters and trusses. I found, intriguingly, the nest entrance total was 15 in the roof span of 5m x 5m, (see photos below).

I got down very slowly and carefully and with a sigh of relief that my vertigo did not act up, we were ready to move on to our next destination. NAP said the owner phoned her and said that he was on the way to meet us. So, we waited a while and looked around the 8.5-hectare farm. When we exceeded the time schedule, we apologized that we could not wait for the owner any longer and moved on.

Figure 162 Top row: Climbing the water tower to examine SB nest entrances in the rafters and trusses totalling 15 entrances. Mid and bottom photos: Splitting an SB colony in a PVC pipe and transferring it into the bamboo.

Figure 163 the farm structures and SB nest in the rafters. Closeup of Tetragonula iridipennis in a roof beam

9.7.4 Closeups of Nest Entrances at Rafters and Trusses of the Water Tank Tower

Figure 164 Closeups of nest entrances by the rafters of the water tower roof. The water tank tower is about 12 meters from the ground and the rafters are another 3 meters

9.7.5 Closeups of *Tetragonula*[24] nest and the caste in PVC pipe

Figure 165 Tetragonula nest in the blue PVC pipe, Top left: isolated worker; Top right: Queen; Bottom left: Brood cells; Bottom right: callow

[24] It was difficult to tell the species but after getting an image of the callow and the pigmentation was very close to that of a *T. laeviceps* callow.

9.8 Kamia Bay Resort

9.8.1 Getting there: Kamia Bay resort is 40 min. northwards along the coastal road from Magsaysay or 40 min. around the Bay from Puerto Princesa

9.8.2 Etymology: Kamia A submission from the Philippines says the name Kamia means "flower" and is of English origin.

9.8.3 Kamia Bay Resort is a water adventure theme park. We have had this destination in mind because a friend contributed a photo of a nest entrance on the trunk of a century-old *Cycas rumphii*, near the entrance of station 4 of the park. Unfortunately, not knowing the business hours, we found it closed on a Monday. We could not venture further. As Christ Mark Bergonio told us he found only one nest entrance there and he had provided us with the photo, we just proceeded to the next destination for our lunch. We were there on March 9th, a Monday.

9.8.4 Resort Entrance[25]

Image 22 Century-old Cycas rumphii with SB nest at the base, the propolis has a varnish aroma. (Christ Mark Bergonio – Kamia Bay Resort)

[25] This location was alerted to me by Christ Mark Bergonio from Cavite, who was here couple of months back. This photo is contributed by him, taken during his family vacation then.

Chapter 10 – Marinduque Island

10.0 Addendum - Meliponiculture activities in Marinduque

Image 23 Marinduque Stingless Beekeepers Association emblem

10.1 Marinduque Stingless Beekeepers Association (https://www.facebook.com/Marinduque-Stingless-Beekeepers-Association-736114896585871/) MSBA is under the auspices of the Federation of Marinduque Agriculture Cooperatives (FEMACO) with the assistance of NMAP, Norwegian Mission Alliance Philippines.

Permission to include all photos in this chapter was granted via Ms. Reynalyn Pilar from Buenavista, Marinduque. Rheyn AP is a Project Manager at Norwegian Mission Alliance Philippines.

FEMACO or Federation of Marinduque Agriculture Cooperatives is a provincial-level federation of primary agriculture cooperatives. At present they have 26 primary cooperative members coming from the municipalities of Buenavista, Torrijos and Mogpog. Stingless bees are a major enterprise of its primary cooperative members. Honey and propolis are among their products. Aside from those, they are also engaged in the production of turmeric powder, ginger powder, coco sugar, peanut butter, peanut brittle, coco vinegar, fish paste (bagoong) and others. Their major advocacy is to generate income from their enterprises to provide social programs/ services for children and youth in their respective communities.

Their office/ consolidation facility is located at barangay Daykitin, Buenavista, Marinduque.

10.2 MSBA fb group…This is what I have gathered browsing the fb group...

MSBA Members

1. Michael Palermo
2. Ryline Delacruz, Tomjos
3. Apostol DT EM, Tomjos
4. Rheyn AP, Buenavista
5. Edna V. Permejo
6. Jenny Naling Sotto Sajul, Buenavista
7. Daulong Esther
8. Ronalyn Ramento, Tomjos
9. Dante Montiano. Buenavista
10. Mechelle Ojena Layag, Mogpog
11. Vincent Flores

10.3 Photo gallery of Meliponiculture activities among Agriculture Cooperatives

Image 24 Top row and Bottom left: PAGKASAMAKA Agri Coop, Argao, Mogpog; Middle row and bottom left: BIDA Agri Coop @Daykitin, Buenavista

Image 26 MASIKAP agriculture cooperative, Sihi, Buenavista

Image 28 ANIB Agriculture Cooperative, Bagacay, Buenavista

Image 27 CASH agriculture cooperative, Caigangan, Buenavista, Marinduque

Image 25 SAMAPI agriculture cooperative - Sitio Pinagbulugan

Image 30 SAPPAT agriculture cooperative, Timbo, Buenavista, Marinduque

Image 29 SAMAKAP agriculture cooperative, Makawayan, Torrijos, Marinduque

Image 31 SNAP agriculture cooperative, Sibuyao, Torrijos, Marinduque

Image 33 SUMAKA agriculture cooperative, Malibago, Torrijos, Marinduque.

Image 32 Marinduque Stingless Beekeepers Association

10.4 Getting there: From Manila by road to Lucena City. Then from Lucena Port by ferry to Port of Balanacan on the north tip of the island. The modes of transport on Marinduque if one would like to visit those Meliponary are Jeepney and tricycle but one can also hire a private van or car.

10.5 Meliponiculture Activities Chart among Coops in Marinduque.

Chart 4 Meliponiculture Activities among Agriculture Cooperatives in Marinduque

10.6 List of Marinduque Agriculture Cooperatives with Meliponary

1. BIDA Agriculture Cooperative from Daykitin, Buenavista
2. SAMAPI Agriculture Cooperative - Sitio Pinagbulugan, Tungib, Buenavista
3. MASIKAP Agriculture Cooperative, Sihi, Buenavista
4. CASH Agriculture Cooperative, Caigangan, Buenavista,
5. ANIB Agriculture Cooperative, Bagacay, Buenavista
6. SAMAKAP Agriculture Cooperative, Makawayan, Torrijos,
7. SAPPAT Agriculture Cooperative, Timbo, Buenavista,
8. SUMAKA Agriculture Cooperative, Malibago, Torrijos,
9. PAGKASAMAKA Agriculture Cooperative, Argao, Mogpog, Marinduque
10. JAPAGSAKA Agriculture Cooperative, Janagdong, Mogpog, Marinduque
11. KASAMAKA Agriculture Cooperative, Buenavista, Marinduque

List of Figures in Part 2

Part 3 Visayas

Foreword

It was a tough, punishing and backbreaking schedule and I can imagine the navigation of Ferdinand Magellan discovering the Philippines five centuries ago. I'm grateful that he also landed first in Cebu—a very symbolic similarity to when Magellan met Rajah Humabon a Muslim Datu in Cebu. While our meeting was short, I gained a lot of knowledge and was honoured to host him a breakfast. His advocacy of promoting, protecting and developing our endemic SBs in the Visayas is worth gold (Like the offer of Santo Niño as a baptismal gift to Hara Amihan, wife of Cebu's Rajah Humabon). I believe in the same fundamental principles as sir Abu. No foreigner nor Filipino bee scientist has ever done that to help us Visayan SB advocates to promote our own and be aware of the potential of native species in the Visayas. I've seen a lot of greedy beekeepers in my 30 years of experience as a beekeeper. Last year's deliveries of 200 colonies of *Tetragonula biroi* in Tacloban that died, 80 colonies in Tagbilaran with 50% mortality and 50 colonies delivered on an island that all absconded should not be repeated. The introduction of non-native SBs in the Visayas will displace our native stingless bees and compete for available resources in the locality. During the visit of Sir Abu to the Visayas, we realized the richness of diverse species of stingless bees in our islands, which need protection. I suggest we continue this chat group with due respect to the creator to convert this into a network of SB advocates in the 3 regions in the Visayas.

This literary material and discussion group can have the following purpose:

1- Uniting the SB advocates in the Visayas

2- Knowledge Sharing Group

3- Clearing House for Introduction of non-Endemic species of Bees in the 3 Regions

4- Network for identifying SB species and for savings threatened feral.

5- Training network for new advocates. Hope "Sir Abu Magellan" would agree!

Alberto Bartolata, is currently the National Vice President of BEENET Philippines, President of BEENET Visayas and VP of BEES Association

Preface

"Do you reside in the Philippines? How is it that you have identified bees in different parts of my country?" was posed to me when I was doing this tour for my Meliponiculture Atlas Project for ASEAN.

I was shocked and stopped in my tracks momentarily when posed this question out of the blue. "Well, that's the information given by the beekeepers themselves... I don't know how they identify the bees. I just put their location on the maps... if they are incorrect, please help notify me. Thanks." This was the humblest answer that I could think of.

Those charts are the distribution of Meliponiculture practices and activities... they do not chart any specific distribution of any specific bee.

I am doing a Meliponiculture ATLAS of Indo-Malayan stingless bees. It is not only the Philippines I am charting... I chart other countries as well in the ASEAN region.

"For now, I have no reason to doubt this information given by you to be incorrect. But your effort of tracking them down on a Philippine Map is laudable to me - as I'm minded to promoting the propagation of indigenous SBs among my group of "Sierra Madre Agroecologists Agriculture Cooperative - SMAAC". Anyhow, "Thank You" for your work." was the subsequent conversation.

"In any case, I am just curious sir, am I allowed to identify bees only if I happen to reside in the Philippines and where is SMAAC located?"

"Of course not. Anyone is free to do your kind of Charting Work - which I do value." came the answer.

"SMAAC is a new grouping of farmers sited in the uplands of the Sierra Madre Mountain range, encompassing the provinces of Rizal, Laguna and Quezon, which has become easily accessible by the Marikina-Infanta Marcos highway - connecting West-Manila & East-Infanta. We operate in the mid-section of the Sierra Madre Mountain range - facing the Pacific Ocean starting from the North to down South of Luzon Island," said Ernesto C Del Castillo.

Trying to expand the conversation, I ventured, "Have you traced any species specific to the Philippines like *Lepidotrigona palvanica* or *Tetragonula dapitanensis*... endemic to the Philippines? Unlike *T. biroi* which was collected by a Hungarian (named Lajos Biro) in Papua New Guinea and later the species name was Latinised by Friese, an Austrian Scientist, and described by him as *Trigona [now Tetragonula] biroi*.

So *T. biroi* is not endemic nor identified by a Philippine resident." The conversation abruptly ended and I guess I managed to impress, as I aspired to.

From this, I gather and presume that Filipinos that "balikbayan" i.e., return from work abroad, have funds that they brought home and are looking for opportunities to invest in to start a new career, especially with some native products.

Herein evolve the SB newbie wannabees. Stingless beekeeping in the Philippines is rapidly developing into an emerging industry. One has to keep pace!

Many freshies think it is quite elementary because they are advised by friends and they think it is easy to engage in beekeeping as they see their friends can survive with such a robust emerging commodity and activity.

They have a broader outlook because they return with some extra experience in life from seeing a part of the world when they were abroad. The key here and my advice to them is to get it to thrive and not only survive. They need to focus on extreme concentration and dedication because SB keeping demands much care and enthusiasm.

I met quite a several newbies throughout my travels in Davao, Bicol, Luzon n now the Visayas those that were focused on and are regular beekeepers. Sometimes my words touched sensitive areas and I apologized for that, but those words that stuck made for better progress.

Even so, when more than one year ago I planned with Fernando Belmonte Jr. of Montalban, Rizal, Luzon, on this island hopping trip to meet SB keepers and to find new techniques or innovations, he could not find any willing or extrovert enough in the Visayas when he posted our request in the many fb group sites. Five years ago there was hardly any stingless beekeeper in the Visayas... they were all in the Bicol region in southern Luzon... but now I found from my visits that the Visayas has the greatest number of enthusiasts of any region in the Philippines.

That was after communication with the following, through social media chats in similar interest group discussions

@Peneda Gerry and @Calvin Centino from Tacloban, Leyte,

@Nora Alanza Peña (Iloilo, Panay),

@Jwens Visayas Stingless Bee (Silay City, Negros Occ/)

@Alberto V. Bartolata (Cebu) that I gathered enough “beeks” and locations to visit.

In the words of one of those newcomers, “I don't encourage people to venture into beekeeping just for the honey... I would advise them to do research. I was working on my beescape for almost a year and luckily I met sir Jwens (Jomar) and Sir Fred Lozada of Gawahon when I finally decided to acquire my first colony. Preparation and dedication are key. My mind is like a strainer: it keeps the right info. and disregards the irrelevant”. (Romeo Pasuelo Gellada, new Beekeeper in Bacolod, Negris)

We expect the typhoon has passed. According to the flight schedule, I should be in Cebu at 4 a.m. tomorrow. I am keen to observe the bees' behaviour in the aftermath of the typhoon, especially in feral colonies.

Introduction

Distribution of Meliponiculturists in Visayas, Philippines. Jan. 2020

Chart 5 The List of Beekeepers and Enthusiasts in the Visayas, Feb. 2020

1. Alan Cabalfin, Leon, Iloilo, Panay
2. Alberto V. Bartolata, Pangdan, Cebu
3. Apolonio Cabangal Jr., Leon, Iloilo, Panay
4. April Fatima Dira Villnueva, Szn Joaquin, Calbayog, Samar
5. Balag Jomer, Cambacay, Batuan, Bohol
6. Bethel Joy Alanza Peña, Cebu
7. Calvin Centino, Tacloban, Leyte
8. Chi Flores, Manggapuri, BSD, Negros Occ
9. Christian Paul, Carigara, Leyte
10. Cor Uranus, Taculing Bacolod City
11. Dave Rodriguez, Cebu City
12. Dionisio Panilagao, Pangdan, Cebu
13. Dodjie Marquez, Bacolod, Negros
14. Evan Calaor, Bucari, Leon, Iloilo, Panay
15. Emanuel Nisol, GEP-Brgy 11, Victorias City, Negros
16. Francisco & Conie Clarito, Leon, Iloilo, Panay
17. Freddie Lozada, GEP-Brgy 11, Victorias City, Negros Occ.
18. Gary Avila Ayuste, San Vicente, Tunga Leyte
19. Herman Caololan, Awis, Leon, Panay
20. Ivan Paul Pauchano, Leon, Iloilo, Panay
21. Jezrael Vergara, Pusok Lapu-lapu, Cebu City
22. Joe Asis, Guinbirayan, Santa Fe, Tablas, Romblon
23. John Cagalitan, Sitio Balud, Poblacion, Leon, Iloilo
24. Jonji de Jesus, Iloilo, Panay
25. Joshua Lupian, San Guillermo, Palompon, Leyte
26. Jun Caidoy Daya, Julita, Leyte
27. Jwens (Fejumar Gasper) Kapitan Ramon, Silay City, Negros Occ.
28. Lorna Delarosa Yugo, Villaba, Leyte
29. Lucita Calantoc Cablas, Talacuan, Leon, Panay
30. Marc Lyndon Genovana, Jordan, Guimaras
31. Matias Tabasa, Pangdan, Cebu
32. Melvin Guerrero, Matlan, Isabel, Leyte
33. Meredith Flores Arillo, Poblacion, Iloilo, Panay
34. Mike Cabatuan, San Guillermo, Palompon, Leyte
35. Mikee Balais, Tacloban, Leyte

39. Noel Regulacion, Juaton, Ormoc. Leyte
40. Nora Alanza Peña, Leon, Iloilo, Panay
41. November Canieso-Yeo, Bacolod, Negros Occ.
42. Pamela T. Henares, Alangilan, Bacolod, Negros
43. Peneda Gerry, Palo, Leyte
44. Perla Cabalfin, Leon, Iloilo, Panay
45. Ricky Sobesta, GEP-Brgy 11, Victorias City, Negros
46. Rico Pietro, Dumaguete, Negros Oriental
47. Ritche Peter Villaneza, Valencia, Ormoc, Leyte
48. Romeo Pasuelo Gellada, Bacolod, Negros Occ.
49. Roque Villaneza Jr. Valencia, Ormoc, Leyte
50. Ruel Benitez Delantar, Maypangdan, Borongan, Eastern Samar
51. Sencia Alminaza Clarito, Talacuan, Leon. Iloilo, Panay
52. Ta Ta, Biliran, Leyte

Meliponiculture Distribution in Visayas Region VI, VII & VIII

List of Meliponiculture activities in Visayas Region VI, VII & VIII

1. Nora Alanza Peña, Leon Iloilo
2. Herman Caololan, Awis, Leon
3. Ricky Sobesta, Gawahon Eco Park/ Victorias City Negros
4. Freddie Lozada, Gawahon Eco Park, Victorias, Negros Occ.
5. Jun Cagalitan in Sitio Balud, Población, Leon, Iloilo
6. Perla Cabalfin, Leon, Iloilo
7. Jumar Gasper, Silay City, Negros Occ
8. Emanuel Nisol, Gawahon Victorias, Negros Occ
9. Christian Paul, Carigara, Leyte
10. Gary Avila Ayuste, Beengo Farm. Tunga, Leyte
11. Municipal Nursery, Sitio Tabionan, Bucari, Leon, Iloilo
12. Ivan Paul Pauchanom Potolan, Iloilo
13. Ta Ta, Biliran, Leyte
14. Mikee Balais, Tacloban. Leyte
15. Marc Lyndon Genovaña, Santa Teresa, Jordan, Guimaras
16. Romeo Pasuelo Gellada, Bacolod
17. Bobjas Nequia, Miag-ao, Iloilo
18. Farmer Lodi, Sugod, S. Leyte
19. Peneda Gerry, Palo, Leyte
20. Jun Caidoy Daya, Julita, Leyte
21. November Canieso-Yeo, Bacolod, Negros Occ.
22. Aboy's Restaurant, Bacolod

Chapter 11 – Cebu Island

I took the night flight from KL to Cebu with plans to execute the following proposed itinerary.

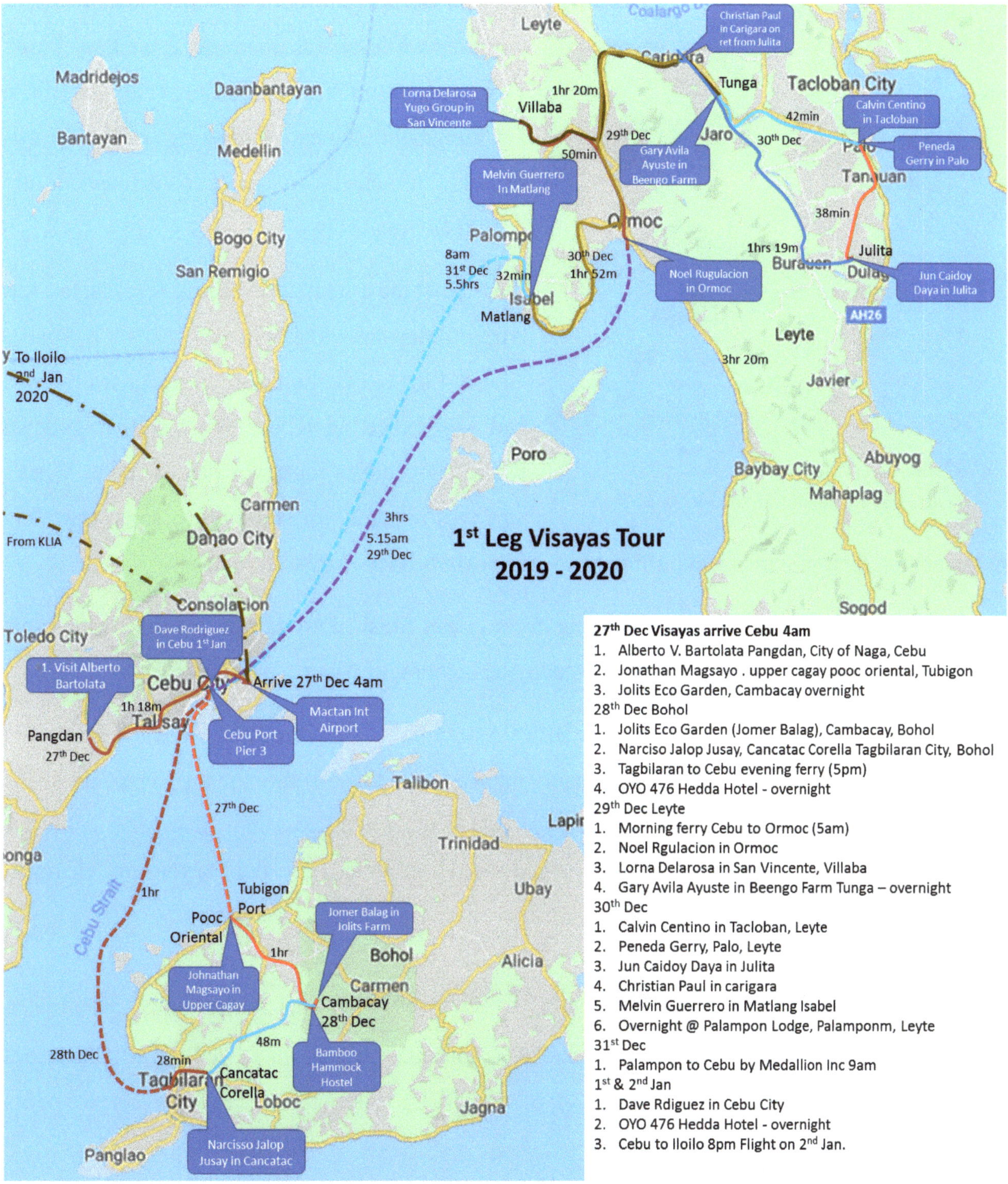

Map 3 Proposed itinerary for the first leg of the Visayas Tour

There was a 2-hour delay in the flight schedule, so the flight took off at 2 a.m. on the 27th and arrived

11.0 Arrival

Figure 166 Arrival at Mactan Int. Airport

The first thing I said to myself was that I needed to get a phone card and Coby was there to help me out with that while he was queueing at the busy kiosk. I excused myself and approach a security guard to ask him about the nearest smoking area. After a 4-hr. flight, it melts down a smoker's demeanour into a child craving candy.

The first thing I learned is that at the Mactan Int. Airport, the smoking area was called a Smoke House. I looked in the direction the guard pointed and was expecting to see some sort of godown where one smokes rubber or fish or copra. What I saw was a gazebo of sorts where some travellers were quenching their cravings and puffing away, I quickly headed there to join them.

Coby was timely enough as I saw him heading my way as I finished my last puff. With phone cards installed, we set out with Jezrael towards Pangdan, the location of Alberto's Meliponary.

Jezrael Vergara is a Beekeeper member of the BEES Association. (Bees for Environment & Economic Sustainability). He is an apprentice pilot and aptly piloted us southwards through the Cebu City traffic toward the city of Naga. My last visit to Cebu was 20 yr. ago when I was an avid seashell collector. The trip was a little more than an hour and we talked about how much Cebu and Mactan have changed.

11.1 Alberto Bartolata – Pangdan, City of Naga, Cebu

Soon we arrived at the home of Alberto Bartolata. He is currently the National Vice President of Beenet Philippines, President of Beenet Visayas and VP of 'The **BEES'** Association. There we met two of his friends who were new stingless beekeepers.

There was Dionisio Panilagao who is a mellifera beekeeper but also raised 10 colonies of stingless bees. He had been working abroad for the past 20 years but retired as a beekeeper. Another was Matias

Tabasa who is a seaman— a mellifera beekeeper and lately acquired 20 stingless bees and raised two feral colonies.

Figure 167 from left Alberto Bartolata, Dionisio Panilagao, AHJ, Jezrael Vergara, Matias Tabasa and Nora Alanza Peña

We went around his house while in intense discussion. One could sense his passion for the bees. He probably had about 200 hives just in his compound and around his house.

Interestingly the boxes were coloured differently as it would give the bees a guide for homing on their return flight. The pastel colours used included white, green, blue, light red and some natural wood finish. His feral hives are bamboos and are stacked awaiting transfer into boxes. However, the arrangement seemed a bit haphazard.

Figure 168 His feral hives are bamboos and are stacked awaiting transfer into boxes, I guess.

Figure 169 Interestingly the boxes are coloured differently as it would give the bees a guide for homing on their return flight.

The bottom corner photo shows the stock of boxes, some black some white and some pinkish. This reminds me of beekeepers in Queensland Australia who practice finishing their box hives in assorted colours and patterns.

Figure 170 One box had a peculiar 2nd entrance/exit at the back corner (right photo) while the front main entrance (left photo) was the same as the others. Opening the top (middle photo) we saw that the exit originates from under the plastic observation sheet meaning it's the same colony.

One box had a peculiar 2nd entrance/exit at the back corner (right photo) while the front main entrance (left photo) was the same as the others. Opening the top (middle photo) we saw that the exit originates from under the plastic observation sheet, meaning it's the same colony.

Issues we talked about were the bamboo hives arrangement and that the distance between boxes should provide a comfort zone clearance, especially during the swarming[26] season. The Beescape also has much that can be added, more so at the groundcovers level.

Figure 171 Left: Co-author NAP with Jazrael, Middle Author AHJ in discussion mode, right: Alberto about to open the top of the hive with the auxiliary exit.

[26] See Appendix 7 on Types of SB swarms

Chapter 12 Bohol Island

12.0 Tubigon

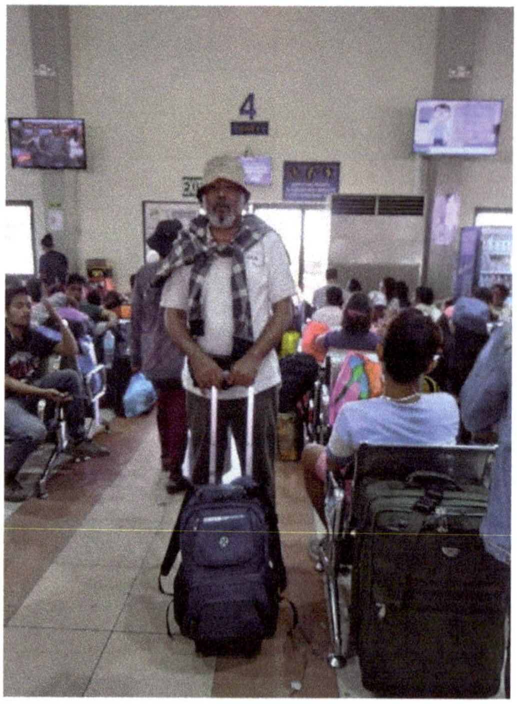

Fig 2 Ferry terminal waiting lounge in Cebu

12.0.1 Getting there: Tubigon is a Port city in Bohol. It serves mainly passengers from Cebu Port. It was sort of a weird experience. I saw the tickets not having seat numbers and I just presumed that was the norm. We were early and it appeared as a regular waiting lounge as it was after Christmas I expected many would be travelling back to each home island. Ma'am Nora had bought the tickets a couple of days prior. We were not too worried about seat availability as we soon saw the ticket lines were bursting.

My tummy started rumbling because we did not have breakfast yet. I looked around for something to munch… not crackers again? I thought this was a large terminal and Cebu is quite highly populated with a fair Muslim population (though a minority) so definitely, they would have some sort of halal food. Seeing none to my fancy, I settled for some plain instant noodles with squid rolls.

Then as it came closer to time to sail, the crowd started to pour in. The waiting lounge soon became a crowded packed bustle of people. As we boarded the ferry, I was glancing at other passengers' tickets and they had seat numbers written in marker ink. As we boarded we were not shown to our seats, instead, we were asked to wait aside. My knees were already wobbly while other passengers were gayly adjusting their bums on their allocated seats. The group without seat numbers was accumulating while blocking the entry into the passenger deck. All those that bought tickets on earlier bookings the previous day did not have seat numbers on their tickets. There was some confusion a few days before as some ferry trips were cancelled due to the typhoon, as such, all those that missed out came back the next day to buy the next available dates. This led to quick ticket issuance without seat numbers and that's where we ended up. Anyway, all's well that ends well. We got our seats eventually (I don't know what system they used to allocate our seats). We were on our way to Tubigon.

12.0.2 **Etymology**: Tubigon in Filipino means water. Origins are thought to be early Malays from southeast Asian countries, migrating to the Philippines in small bangkas. Seasonal overflowing of the river flooded the community and led the place to be called "Tubigan" (meaning "place having water" or "watery"), which later morphed into "Tubigon" (meaning "a place abounding in the water"). The first record of the town of Tubigon dates from the early part of the 17th century.

12.1 Nathan Amizola Magsayo in Pooc Oriental

He goes by different names, he addresses himself as Jon in messenger chats, and hence the name Jonathan Magsayo makes sense although his alter-ego prefers the nickname "Wakooh Ehl Hee" on Facebook. Well, to each his own, and we headed on to Jon's home in Pooc Oriental. Early in the

Figure 172 Magsayo residence in Tubigon

afternoon, we were on our way to the Wessam Express to get to the water and safe to get to Tubigon Port. Jonathan Magsayo, who was known as Wakooh Ehl Hee. He met us on Facebook and had a convoy to their house while we were on the tricycle. He showed us his pet bees in front of their house and on the side near the garden.

He builds boxes in the typical shoebox style although they seem to be for size 12 shoes. The boxes hung under roof eaves and a couple of bamboo hives. The species appeared mostly to be *T. laeviceps*.

We talk about doing swarm traps, proper beescaping, hanging bees to avoid predators and how to increase the colony they care about. We left Jonathan's home smiling because of the immense knowledge he had shared that afternoon. We took a tricycle back to the Van terminal just near Tubigon Port.

Figure 173 Boxes hung under roof eaves and a couple of bamboo hives. They appear mostly to be T. laeviceps

We left Tubigon past five o'clock in the afternoon.

12.2 Batuan -Jolit's Eco Garden

12.2.1 Getting there: From Tubigon as private taxis are not available, public vans ply the routes to the interior of Bohol. We boarded a van and headed to Carmen. Jomer Balag was waiting in the town centre and from there he guided us to his garden in Batuan. We all rode on motorcycles to his place.

12.2.2 **Etymology**: Batuan in Filipino means rock or rockery. It amazes me that we were going to a location named for its rockery to visit an eco-garden. The barangay is Cambacay and I was already booked at Batuan Hammock Hostel which is a dormitory-type accommodation, quite suitable for Ma'am Nora, her son Coby and myself.

This location was not far from Jomer's place. Again, here I was amazed looking at the photos when I booked. The landscape was lush for a name called Batuan. Jomer insisted I cancel that booking and stay at his place because he just completed a couple of chalets for guests.

12.2.3 **Accommodation**: The chalet at Jomer's Garden was a pleasant surprise. We were initially prepared to rough it out, but this was roughing it out in multiple-star Permaculture luxury.

Figure 174 The chalets at Jomer's Garden were a pleasant surprise. We were initially prepared to rough it out, but this was roughing it out in multiple-star Permaculture luxury.

12.2.4 **Perma-Culture** derived from Permanent agriculture will reflect activities of sustainability and recycling of waste. True to his concept Jomer Balag practices Permaculture to the fullest extent compared to any other setups that I have ever seen. He has ducks, geese, chickens, tortoises, and rabbits among herbs, vegetables and fruit trees. The other, ruminant husbandry, is in another location on his large farm.

Figure 175 Innovative examples of hive constructions

Figure 176 Jolit's Eco-Garden in its complete Permaculture glory

12.2.5 **Vermiculture & EM:**

Figure 177 Variety of nest entrances, all appear to be Tetragonula spp.

Jomer is quite an avid vermiculture farmer himself besides dabbling in agricultural EM production. While collecting garden snails and earthworms he also rears mealworms. To the garden snails with compost and molasses, he adds propolis and produces effective microbial communities. The earthworms and mealworms leave vermicasts and mealworm casts. Nothing goes to waste. I am so excited looking at his whole system working like clockwork.

12.2.6 Meliponiculture:

As a Beekeeper, Jomer is continuously experimenting and innovating. He keeps bees in wooden cuboid boxes and even in terracotta clay pots. I found that he has almost all clustered blood cell bee species, but at least five different structures and morphs.

Figure 178 Wooden box hives and other types

Figure 179 Brood arrangement in Poky styrene box

12.2.7 **Garden of Jomer Balag by NAP**

At four in the afternoon, we met Mr. Jomer Balag in Carmen town and then we took a tricycle to Jolit's Eco Garden. As we arrived, even in the midst of darkness and strained eyes, we were amazed at the beauty of the place. Jomer took us to his new homestead to put out our luggage and lock our cellphones. We especially had the urge to cross the small bridge to the lodge on the lake made by Jomer. Also, pay attention to his bee sitting in a large Styrofoam box. A long discussion that night when all the information from an Agronomist (Coach Abu) and Agriculture Scientist (Jomer) came together.

In the morning, it was still dark but before long we were awake with the owner of this garden. It was not before the sun appeared that we started walking around the garden again and we noticed that Jomer was cleaning the area. One by one we photographed each part of the garden and did not miss the beehives and opened them to determine their breed.

I was fascinated by the size of the hybrid passion fruit, smooth, flowering, and large fruits of Cacao flowering plants. I also picked the ripe fruit of the native passion fruit to harvest the seeds. Jomer is adding more to my crop. We went to Jomer's guest house again to have breakfast. He introduced us to his wife, Mrs. Lita Balag. We realized that Jomer had no staff to help him maintain the beauty of the garden but had the help of his children. At ten o'clock in the morning, Jomer took us to the town of Batuan to get transportation to Corella our next destination.

12.3 Brgy Cancatac, Corella, Bohol

Corella is known primarily as the home of the endemic Philippine tarsier, one of the world's smallest primates. Its former name was Nug–as before it obtained its present name, Corella. It was then a barrio of Baclayon. The adoption of the name Corella was made at the behest of Fr. Jose Maria Cabañas del Carmen, then the parish priest of Baclayon and endorsed by Fr. Felix Guillen de San Jose, the first Spanish priest. Corella was a name of a village in Navarra, Spain where this town's patroness, Nuestra Señora del Villar, showered miracles.

12.3.1 Getting there: From Tagbilaran, we took the bus to Corella and got down at the spot Narcisso had indicated, where he was waiting with a friend. We rode on the motorcycles with me behind Narcisso and Ma'am Nora and her son Coby on the friend's bike. I managed to snap some photos over my shoulder along the way. It was a dirt road for a couple of km. to reach his village.

Figure 180 Along the way to Caracan from Corella

Coby, in turn, took a photo of me riding on the bike in front. We arrived at his village, parked the bikes and had to walk a short distance to Narcisso's house.

While waiting in his house I noticed Narcisso has some pretty and unique antique chairs. He said they belonged to his wife's grandparents. The carvings were very similar to Malay floral carving motifs. Anyway, after some vegetarian lunch, I went down with him to check out his bees on the stilts under his house.

12.3.2 Meliponiculture: He had some nests in the house joists and on the stilts. He had a bamboo hive and one coconut shell stacked hive, this, however, does not seem to be active. The one in the box looks like a former *Apis* hive and is now with SBs.

Figure 181 Top left: Coby in turn took a photo of me riding on the motor in front. Top right arrived at his village and they parked the motorcycle a distance away/ Bottom left walking towards Narcisso's house. Bottom right: the house where they parked the bikes.

Figure 183 assorted SB in the house structure and he has a bamboo hive which appears to be the result of a swarm trap and a box hive which looks like a former Apis sp. hive.

Figure 183 some closeups of nest entrances at Narcisso's house in Cancatac, Corella, Bohol

12.3.3 Cohabitation of SB in Termite infested tree. a few hundred meters behind his house we saw a tree infested with termites and just waiting to fall. I said that at the next typhoon with the right wind velocity, he will get a new SB colony.

Figure 184 Cohabitation of SBs in termite-infested tree

Chapter 13 Leyte Island

13.1 Ormoc Port

13.1.1 **Etymology**: The city's name is derived from *ogmok*, an archaic Visayan term for "lowland" or "depressed plain".

13.1.1 Getting there: From Cebu port onboard the 6 a.m. ferry we arrived at Ormoc port at 9 a.m. As we were looking for the person to fetch us, Ma'am Nora was texting the guy mentioning our location and suddenly someone emerged and introduced himself as Noel. With much relief, we walked with him to the nearby terminal to catch a Jeepney to go to his Barangay.

13.2 Noel Regulacion, Brgy. Juaton, Ormoc City

Noel Regulacion was referred to us by Lorna Dela Rosa of Villaba. She wants to set up a bee community there. Initially, Noel was supposed to guide us to Villaba. In the Jeepney, which was a tight squeeze, (one doesn't realize how fat one is until one sits in a jeepney with equally chubby ladies going home from the market), I was wondering, are we going all the way to Villava in this? The ladies in the Jeepney were all staring at me and I was wondering, don't I look Filipino enough with my dark skin… hmmm, must be my macho beard. Anyhow, Ma'am Nora mentions something in Visayan and they all went "Aaah!" In Leyte, a couple of dialects are spoken namely Cebuano and Waray, none of which I understand a single word of. Well, that's why Ma'am Nora is around as an interpreter.

13.2.1 Feral colonies: Plans changed, however, and we ended up visiting Noel's barangay of Juaton. He brought us to his house and showed us some feral colonies existing around his house. He then suggested we visit the Jaton Barangay Hall to check out a colony in the planter that has been there for over six years. Along the way, we met Noel's wife with a few ladies from their community.

Figure 185 visit the Jaton Barangay Hall to check out a colony in the planter that has been there for over 6 yrs. Along the way, we met Noel's wife with a few ladies from their community.

Figure 186 Noel Regulacion with the oversized cocoa pod and a couple of feral colonies at his house in Juaton, Ormoc

13.3 Ritche Peter Villaneza's farm - Brgy Valencia, Ormoc City

Figure 187 Noel plucked a ripe Papaya for us and is such an expert at it.

13.3.1 Getting there: We rode on motorbikes towards Valencia to visit Ritche Peter Villaneza's farm. Along the way, we could see many trees devastated by Typhoon Ursala just that past week. Most of the devastation was at his Jackfruit (Nangka) plot.

Many of the trees had branches broken off due to the high winds. The family farm is producing sugarcane. He later brought us to his home, where we met his father, his brother Rey Philipp L. Villaneza and his wife.

Figure 189 On the way to Ritchie's Farm

Figure 188 Ritche indicates a Bamboo feral hive that has been damaged by rats.

13.3.2 Rats Looting:

Figure 190 An abandoned cabinet with SB nests at every corner. Unfortunately, it has been looted by rats. Ritche hasn't had time to open to inspect the interiors because he is currently refurbishing the farm just after the Typhoon.

13.3.3 At Ritchie Peter Villaneza's farm in Brgy Valencia, Ormoc City.

We hop on the motorcycles of Ritchie and Salvador while Noel drives alone to Ritchie's garden. We immediately headed for Ritchie's beehive. We were again impressed by his generosity and did not hesitate to get pomelo, ripe papaya and bananas to pack on our trip to Beengo Farm. We did not expect that we would be entertained at the luncheon at his home and we also met his wife, brother Rey Philipp L. Villaneza and Father Roque L. Villaneza with Malay blood. It's a fun story of new friends, really different because everyone immediately agreed and exchanged information. We again said goodbye to Ritchie's family. We took a jeep to the Van terminal.

13.4 Tunga, Leyte – Beengo Farm

13.4.1 **Etymology**: One states that according to the forefathers before Tunga became a barrio, people from Ormoc, Carigara, Barugo, Jaro and Tacloban engaged in trade with each other and they had to stop by or pass this place. Some of them even made it their contact point for conducting their business. Through their exchange of ideas and conversations, they concluded that this place was halfway between Ormoc and Tacloban or Carigara and Jaro. Since it had no name at the time, people started calling it Tunga, the dialect for half. (In Malay, halfway or mid-way is "tengah").

13.4.2 **Getting there**: We were told to inform the van driver that we wanted to go to Beengo Farm and he will tell us where to get down. He promptly did so and there was a roadside makeshift shed, where I suppose people would wait for tricycles. We got one and told him our destination, the guy went "huh?" Ma'am Nora spoke something in their dialect I guess, and then the guy went "Aaah!" and nodded. This reminds me, if ever I repeat this trip in the Visayas, make sure I have an interpreter. There are just too many and varied dialects. The dialect in the region here is Waray. And when you have an interpreter, keep your mouth shut and behave like the senior citizen you are who can't be bothered to talk.

13.4.3 Beengo Farm – Gary Avila Ayuste

Figure 191 BEEngo Farm façade

Figure 192 The Hobbit House With A Green Roof

Figure 193 Riverside huts adding to the "Nature Ambience"

13.4.3.1 Ambience:

The elements Gary created and arranged captured the Nature Ambience he was aiming for.

13.4.4 Meliponiculture: Gary has a collection of box hives brought in from a few locations. Most of his *T. biroi* is from Luzon and includes one *T. laeviceps* from San Guillermo, Palompon. However,

Figure 194 One from Luzon and one from San Guillermo, Palompon

during the last typhoon just last week, he was badly hit. He is reconciling them on one of his riverside huts.

13.4.4.1 Novelty Hives:

Figure 195 Assortment of Novelty Hives

13.4.4.2 Chapel SB Colony: Gary suggested I visit a Chapel nearby to view some peculiar SB nests

Figure 196 He showed a nest entrance that had white particles at the cerumen in the Chapel pillar.

there. He showed a nest entrance that had white particles at the cerumen. The white particles did not occur when he brought them back to the farm (after eduction). I told him that there is a high possibility that the bees were attracted to the white glossy paint used on the pillar.

13.4.4.3 School Wall Bee Mural: On the way back to the farm from the Chapel, we stopped at the Banawang Elementary School to view some wall paintings done by the students.

Figure 197 Banawang Elementary School Bee Mural - some wall paintings done by the students.

13.4.5 Getting to Beengo Farm, Tunga; by NAP

We drove the Ormoc City terminal van to Tacloban City. We were very lucky on this trip because it was continuous and without any obstacles. We reached Beengo Farm at 3:55 p.m. From the alley, we took a tricycle to Beengo Farm.

The arrangement of the plants seems like a paradise, the bees being housed in a variety of beehive materials. The location of the place is nice and the house is well-designed to make guests feel comfortable.

13.4.6 Addendum BEEngo Farm; by Gary Ayuste

Description and History

Beengo Farm is a BEE and Mango Farm located in the smallest town by area and by population in the entire Eastern Visayas, Tunga, Leyte. Owned and managed by Gary Ayuste better known as "Mr. Garybee" who used to work abroad for a couple of years, going back abroad after his short vacation in Leyte to visit his family, he missed his flight not just once but twice in a row. His father told me there was a reason why and advised him to stay longer. During his extended vacation, he decided to kill the boredom by just gardening, surprisingly, three months later, people got to notice and like his garden until they were attached to his small garden by the public, and that is how BEEngo Farm was made. To make the farm educational, the owner Mr. Garybee held beekeeping workshops at Milea Bee Farm in Batangas. Garybee also enrolled in the Intensive Beekeeping Course under the UPLB Bee Program at the University of the Philippines Los Baños, Laguna.

We say BEEngo Farm is not just an ordinary place to visit and enjoy, it's a place where you learn and enjoy at the same time. It's a sanctuary for our native honey-producing bees, especially for the stingless bees and *Apis cerana* or ligwan. Since the owner knows that bees are the lifeblood of the food chain, and 85% of plants and trees exist because of the bees, he started his beekeeping career and was able to create a sustainable community. Visitors to BEEngo farm will enjoy foods that are organically prepared with no chemicals and frozen items, and the majority of the supplies are coming from the local community.

The main goals and advocacy of the farm are to save and keep bees not mainly for the honey but to help the farmers increase the harvest of their crops. BEEngo farm also provides free lectures to all visitors on the importance of bees in terms of pollination enhancement and improving biodiversity. BEEngo farm is now a certified Learning Site of the Agricultural Training Institute, ATI, for diverse farming. And we also organized to launch the very first bee-friendly campus in Leyte, the Banawang Elementary School. Since then it has been our mission to erase the stigma from when we say bees, as people are scared because of their stinging. We provide an info caravan on how to properly care for bees without harming you and we also introduce stingless bees and highlight them as friendly and safe for the kids to keep. We donated colonies to some schools to encourage the youth and make them appreciate beekeeping and it is pretty much effective.

To fulfil the owner's dream of developing a sustainable community, he organized a group of farmers who are now called the Empowered Farmers Association of San Vicente. Members are mainly the mothers in the community to create woman empowerment in the barangay. BEEngo farm also extended this advocacy to other places like Pagadian City in Mindanao where a series of workshops was conducted, and also in Calbayog Samar and San Vicente Palawan. BEEngo Farm and Mr Garybee will never stop and will always continue POLLINATING GOODNESS!!! Let us all act like a BEE!

Image 34 Empowered Farmers Association of San Vicente

Image 35 Certified learning site of Agricultural Training Institute or ATI for diverse farming

13.5 Tacloban City - Mikee Balais

13.5.1 **Etymology**: The name came from being a favourite haunt of fishermen. They would use a bamboo contraption called a "taklub" to catch crabs, shrimps or fish. When asked where they were going, the fishermen would answer, "(to) tarakluban ", which meant the place where they used the device to catch these marine resources. Eventually, the name Tarakluban or Tacloban took prominence.

13.5.2 **Getting there**: We took the van from the main road at Tunga and headed for Tacloban, arriving at the city we got down and texted Mikee our location. He met us on his motorbike, and we hailed a tricycle and followed him to his house. There we met his aunty who showed us some photos showing how badly hit they were during the recent typhoon.

Figure 198 Mikee was badly hit by the typhoon as shown by his aunty Fely

13.5.3 Meliponiculture: Mikee gets his hives from sources outside Tacloban City. A couple of intriguing colonies were one from Biliran Island as he declared and one nested in a fern tree trunk. He has been experimenting quite a bit. one colony was in PVC pipe even. While we were there, a parcel arrived and it contained frames for an *Apis* hive that he imported from abroad.

Figure 199 A fern hive that he added a topping box, ideally to keep the mother hive intact

Figure 200 Nest entrances of different species. Left photo shows how that particular species picks different paint particles due to a lack of vegetable resin sources in an urban environment.

Figure 202 Mikee shows us a hive he acquired from Biliran Island. I managed to catch a moment with its queen.

Figure 201 While we were at Mikee's, Peneda Gerry arrived from Palo and join in our discussions on meliponiculture.

Comparing broods, *T. biroi* from Luzon (left) and layered brood from Biliran Island (right). We made some comparisons of the brood and nest structure. The initial differences we observed were that the brood on the left one was more concave and had spiral tendencies whereas the one on the right was flatter in appearance. Secondly, the involucrum on the left was softer in texture and more enveloping whereas the one on the right was more open and stiffer. These were preliminary notes and require more detailed studies.

Figure 203 Comparing brood nests, T. biroi from Luzon (left) and layered brood from Biliran Island (right)

Mikee opened his PVC pipe hive and then I saw how fungus-infested it was. I was compelled to help him transfer the whole colony into a new box hive. Before the transfer, I showed how to prepare the new box by treating it with some propolis extract. The only solvent he had was paint-thinner, so I showed him how to dissolve propolis and then dry it to reduce excess and the thinner aroma.

Figure 204 Mikee opened his PVC pipe hive and then I saw how fungus-infested it was. I was compelled to help him transfer the whole colony into a new box hive. Prior to the transfer, I showed how to prepare the new box by treating it with some propolis extract. The only solvent he had was 'paint thinner', so I showed him how to dissolve propolis and then dry it to reduce excess and the thinner aroma.

13.5.1 Mikee Balais in Tacloban City

It was early in the morning when we reached Robinson Marasbaras, Tacloban City. Mikee came out to take us home. He offered us his prepared breakfast of a bun sandwich, coffee, suman latik and banana chips. We also met his kind aunt Fely Sommer.... After resting a little from the hour's drive, Mikee showed us his pet colony of *Tetragonula sapiens*. We were even more impressed when we saw the colony of the still-extinct Biliran. A colony's nest entrance was in blue, white and red which caught our attention and became a big question for Mikee. Coach Abu explained very well why it is so colourful. This is because of Mikee's house paint. Mikee took us to the back of the house where his colony was located in the PVC pipe. The beehive was opened for us to take a closer look, and to our great surprise when we saw the danger to the colony due to the mould creeping around the brood. Coach Abu did not hesitate to show us the proper process of moving the colony to a beehive. In this part, we were joined by Engr Gerry Peneda. We kindly said goodbye to Mikee and her aunt who held the beautiful Saint Simon flower as a token of our friendship.

13.6 Palo, Leyte - Gerry Peneda ng Mabolo

13.6.1 Engr. Gerry Peneda ng Mabolo, San Fernando, Palo, Leyte

13.6.2 **Getting there:** From Mikee's place, Engr Peneda drove us towards his home in Palo. As it was almost 1 p.m. and we were all hungry after the hard work at Mikee's place, we stopped along the way at a decent Seafood Restaurant called Tia Tita's for about 25 min. away. The spread was quite sumptuous, and I had my fill for the day. I knew it was going to be an arduous journey ahead. Almost finishing lunch, Jun Caidoy was texting Ma'am Nora asking where we were. Telling him we were finishing lunch, I decided to have something packed up for him.

13.8.3 Backyard Meliponary. A few minutes later we arrived at Engr. Peneda's house and he showed us his hives in the backyard. I briefed him on how he could improve the vegetation with a brief Beescape tutorial. Also, how the dried leaves of his creepers need to be cleared off to avoid bacteria and fungi.

Figure 205 I was telling him how he could improve the vegetation with a brief Beescape tutorial. Also, how the dried leaves of his creepers need to be cleared to avoid the build-up of bacteria and fungi.

Figure 206 Some of Peneda's hives, although not really doing too well as they were recovering from the previous week's typhoon

We did not have much to observe because his meliponary was recovering from typhoon Ursala which hit north Leyte the previous week. He offered us some refreshments and fruits after which we made our way to the next destination, Julita.

13.7 Julita, Leyte - Jun Caidoy Daya

13.7.1 Jun Caidoy Daya Meliponary

13.7.2 **Getting there**: I was quite pleased that Ir. Peneda did not take too long to get us to Julita. While chatting about bees, in no time we arrived. After meeting Jun's parents, they suggested proceeding to their farm a short distance away.

13.7.3 Farm Meliponary:

The farm was severely hit as most parts of the farm hut/shed were devastated by the typhoon.

Figure 207 This box shows a healthy brood.

Figure 208 Quite sad to see the destruction though some feral colonies were still around.

If you see brood only in one corner or to one side of the box, chances are there is fungus. Look for whitish moulds, this means the moisture in the nest is high or the lumber was not properly dried. If you see beige, grey or light brown moulds, it could be due to humidity seeping in from poor box construction or poor insulation and condensation caused by box walls being too thin.

Figure 211 Feral colony still surviving albeit a little shy; it shows the resilience of the bees even after high winds

Figure 210 Brood only in one corner or to one side of the box; the chances are they have fungus, and the box is too large

Figure 209 Rummaging through the debris we found the fruit of the Artocarpus gomezianus, a good source of resins for nest building.

Quite an exhilarating experience to see what typhoons can result in. The resilience of this beekeeping family must be commended as well.

Figure 212 Opening a bamboo hive containing typical T. laeviceps brood

13.8 Carigara, Leyte - Christian Paul

13.8.1 Christian Paul Residence

13.8.2 **Getting there**: From Julita, Engr. Peneda drove us via Burauen and headed straight for Carigara. It was getting late evening and I was worried we might not have daylight by the time we reached Carigara. We did reach Carigara with the sun setting but we had to ask around for directions and by the time we found Christian's house's exact house location, it was already night.

Although nightfall, the mood was still festive, and most houses still had Christmas lights blinking, including Christian's house.

13.8.3 **Etymology**: Carigara was originally called Kangara, meaning "that of Gara" or simply "Gara's". Gara was said to have come from Borneo, one of the unnamed companions of the ten datus who landed in Panay and purchased that island from the Ati (or Aeta) chief Marikudo. Later, for phonetic convenience, Kangara became Kalgara and when the Spaniards came, they called the place Carigara; hence its present name.

13.8.3 Front yard Meliponary: Christian placed his hives in front of his house amongst potted plants. He had his feral hives in cut logs at the windowsill. Christmas lights were blinking for the bees too.

Figure 213 A couple of feral hives outside his windowsill

Figure 214 His boxed colony was retreating because of fungus and also due to oversized dimensions

All said and done, Christian accompanied us to the highway to wait for a van to Ormoc.

13.9 Isabel, Leyte - Melvin Guerrero

13.9.1 Melvin Guerrero of Matlang

13.9.2 **Getting there**: We got the van and arrived at Ormoc and sought another van headed for Isabel but Melvin told us to get off at Matlang. Melvin was waiting for us at the roadside with a couple of motorcycles to bring us to his parent's house. They had prepared some freshwater prawns (those types with one larger pincer) they called "takla", fish and vegetables for dinner.

13.9.1 Melvin Guerrero of Matlang

Figure 215 1864 gold coins bearing the name Isabel

Map 4 Route to Matlang from Carigara

3.14.3 **Etymology**: The name Isabel is a matronymic dedication to Isabel II of Spain as evident in the gold coin from 1864. The reigning queen of Spain at the time was minted in 1856.

13.9.4 Meliponary: The next morning Melvin was ready to bring us to his own house. Before that, we had a chat about the village activities with his father and the type of vegetation commonly found in that barangay. As it turns out, Melvin serves as Agriculture Extension Officer in the Agriculture Dept. in Palompon. No wonder he is well versed with botanical names of the plants, I thought.

Figure 216 In the father's front garden with Melvin and his father.

We arrived at his place shortly. He had a fairly large garden and the hives were in the side garden and backyard.

He opened a wooden box and I saw the rarely seen involucrum of a clustered brood, indicating something may be amiss, possibly fungus or pests or contamination (Fig. 199).

13.9.5 Wooden box hives

His boxes had some fungus issues as evident on the inside walls and the covering of white moulds. The weather may be the cause, but steps need to be taken to improve the box construction. He was also lacking a plastic sheeting cover to prevent moisture from entering the box (Fig. 207).

Figure 217 The rarely seen involucrum of a clustered brood indicates something may be amiss, possibly fungus or pests or contamination.

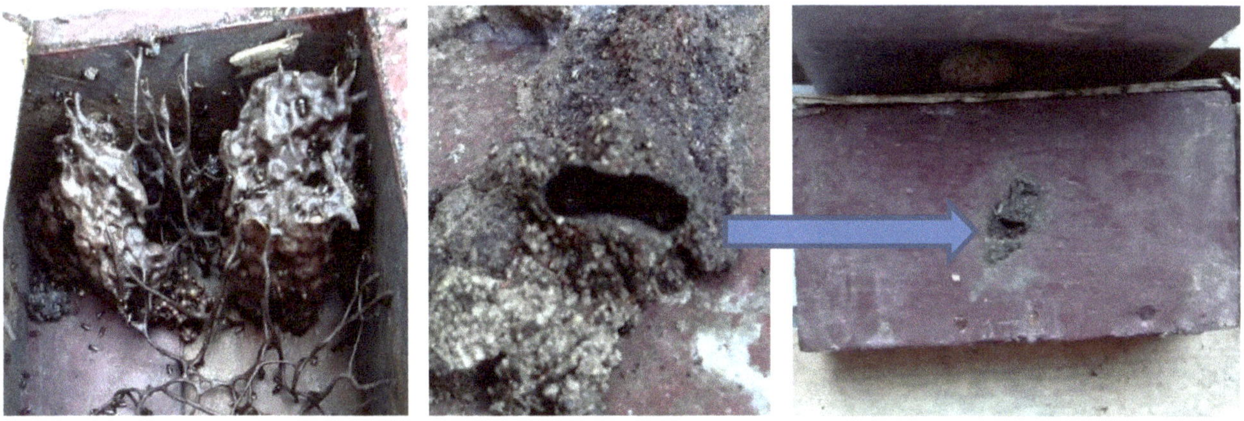

Figure 218 Getting ready to proceed to his own house, he has a fairly large garden and the hives were in the side garden and backyard

13.9.6 Terracotta Urns:

Figure 219 Tetragonula cf sapiens colonies in terracotta urns suspected to be thriving for 10yrs.

Melvin has 2 large terracotta urns that originally belonged to a neighbour. He said that the urns were just lying there and one day they found that bees had made their nests some 10 yr. ago. They like the conditions in the terracotta urns and have been thriving there since. The photo on the bottom left is a close-up of the brood and the bees are nonstressed and docile. He just acquired this colony months ago. He has to configure a sustainable method or continuous harvesting of honey, propolis and beebread.

13.9.7 Odd container hives & Portable Bamboo Hive:

Figure 220 Modified bamboo hive: portable and can be moved in for pollination services

Figure 221 Melvin found some hives in odd places and odd containers. One is in the cavity of an upturned wooden through and another in an empty black plastic container

13.9.8 Beekeeping family:

Figure 224 After a hefty breakfast with seafood partly contributed by Joshua Lupian from Palompon, we bid goodbye to his wife and children with a group photo

Figure 223 Queen in the urn; Brood in Urn 1, Brood in Urn 2

Figure 222 Colony in the mortar through cavity

13.10 Palompon, Leyte – Joshua in San Guillermo

13.10.1 San Guillermo, Palompon

13.10.2 **Getting there**: We travelled by motorcycle as suggested by Melvin because he plies that route daily to go to work in Palampon. Before reaching Palampon Port we diverted into a road leading to San Gullermo.

13.10.3 Etymology: The Barangay is named after the miraculous patron Saint San Guillermo (St. William).

The legend says that when the Spaniards came, they saw floating at the bay a cluster (pong pong) of mangrove propagules locally known as "Tungki". They decided to change the name of Hinablayan to Paungpung, after the cluster of mangrove propagules to erase the bloody memory of the Moro raiders. Gradually the name evolved to Palompong, then to Palumpun, and its current spelling of Palompon sometime in 1700 or 1800. The word pelampong refers to a floating lifesaving device in Malay.

13.10.4 Sea Cucumbers: San Guillermo is a fishing village and the main commodity is sea cucumbers.

The suggested maceration recipe in honey is practised in Langkawi Island, Malaysia.

The Sea Cucumber (Gamat, *Actinopyga* sp.) is cleaned, and the stomach is removed, then cleaned and placed in honey. After 2 weeks it is chopped and then put back into honey and can be used for medicinal purposes or as a supplement. The most popular use here concerns spinal disorders, waist problems, urinary tract, and lethargy. The remedial effects on the body are very noticeable after taking it, especially the symptoms of swelling or inflammatory, which means an increasing the metabolism rate of the body.

Figure 225 His boxes have some fungus issues as observed on the inside walls and the covering of white moulds. The weather may be the cause but steps need to be taken to improve the box construct. He is also lacking a plastic sheeting cover to prevent moisture from entering the box

Figure 226 Sea cucumbers of the fishing village Guillermo are dried salted and sold to Chinese medicine shops in Cebu City

The small black sea cucumber is called “Putian“, the middle one is called “meroy“ and the biggest coloured light brown and the other coloured black are called “hanginan“, and the brown called tiger-tiger is rare and expensive.

13.10.5 Meliponiculture

Figure 227 Most of Joshua's hive interiors had fungus issues and are mostly a bit too large for this Tetragonula species

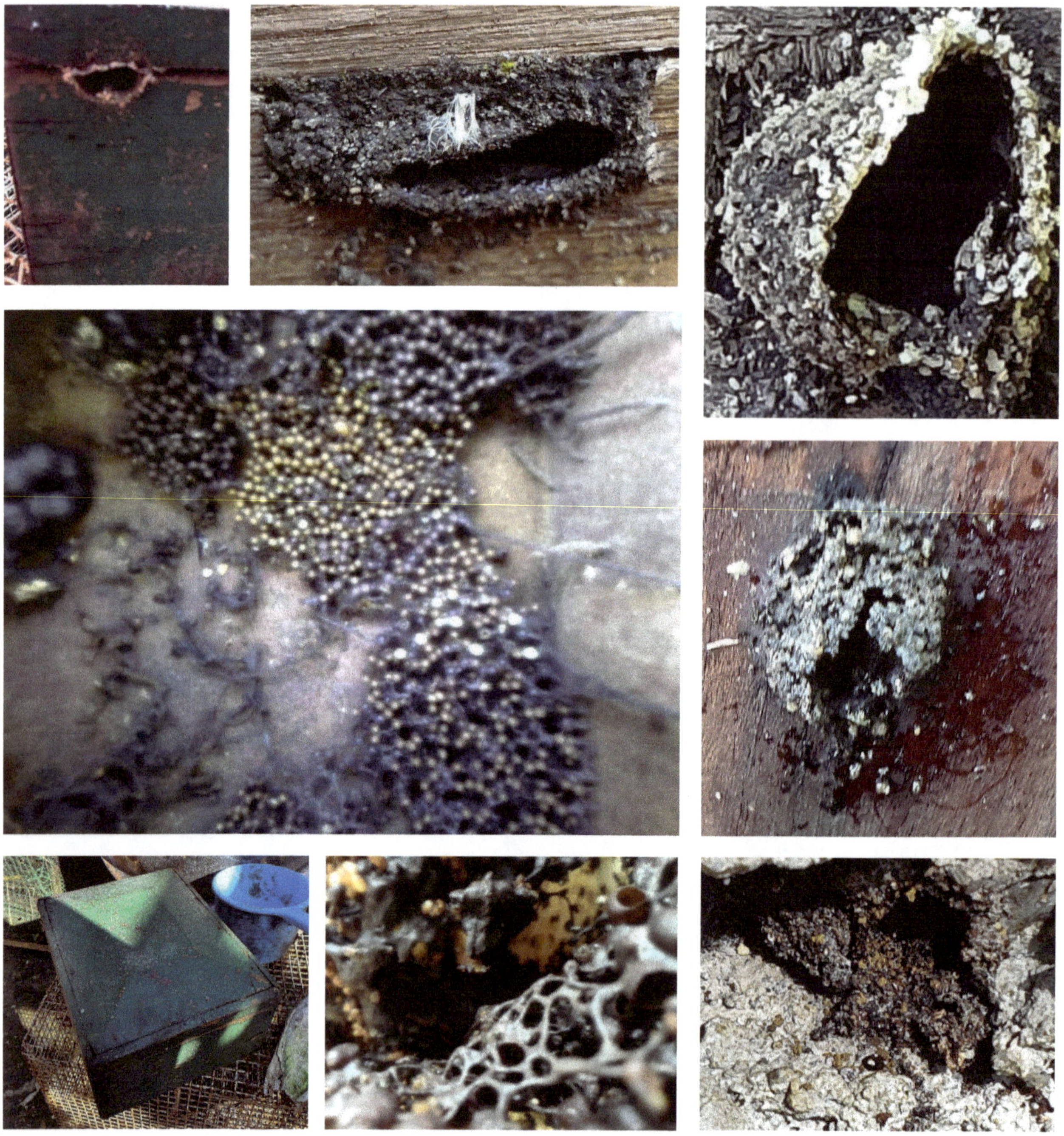

Figure 228 The varied nest entrances indicate there are at least four species of Tetragonula

Figure 229 His backyard may need some rearrangement, but then again, they just underwent a bad hit from Typhoon

Most of Joshua's hive interiors had fungus issues and were mostly a bit too large for his *Tetragonula* species. The varied nest entrances indicate there are at least four species of *Tetragonula*. His backyard may need some rearrangement, but then again, they just underwent a bad hit from the typhoon.

13.11 Sugod, South Leyte - Lodi Carbonilla

Interviews with Lodi Carbonilla aka Farmer Lodi in Sugod, South Leyte.

Q1: Where is the source of your bees?

Ans: Just around the farm.

Q2. How long do you have bees?

Ans: Month old, I have 6 colonies now.

Q3. What design of hives are you using?

Ans3: Wood box 4x4x8 inches

Q4. As of this time what species do you have and how many?

Ans4: SB colonies ..can't identify the species

Q5. Name of the Farm and products produced.?

Ans: The name of the farm is #kdif #, King David Integrated Farm

13.11.1 **Getting there:** It is about 2.5 hr. from Ormoc airport or 15mins from Port of Sogod

13.11.2 **Etymology**: Filipino meanings for sogod are: dash noun 90% of use; large noun 9% of use Cebuano meaning of sugod- commence, n. beginning, n. start

13.11.3 Integrated Farm:

Image 36 King David Integrated Farm (#KDIF) Surrounded by coconut palms they have Celosia sp., Annuals and Sunflowers

13.11.4: Meliponines:

Image 37 A Feral colony in a log transferred into box hive

Image 38 some peculiar Tetragonula sp. nest entrances encountered sporadically

Chapter 14 – Negros Island

14.1 Gawahon Eco Park

14.1.1 **Getting there**: We arrived in Bacolod Port and were told to take a Bus to Victorias City. Of course, to take a bus, one has to take a taxi or tricycle to the Bacolod Bus Terminal. We asked for a couple of tricycles and they were quoting prices equal to that of a taxi, so we opted for a taxi. In the taxi, we negotiated if we were to hire him direct to Victorias. I estimated that we managed to negotiate and probably got about 25% less. We decided to go for it because it was evening and we were losing daylight.

14.1.2 **Etymology**: "Gawahon" is a Hiligaynon word that means "overlooking". Concealed in the abundant greens of the eco-park are the magnificent waterfalls of Gawahon.

14.1.3 Eco Park *Eucalyptus simulacra*:

Figure 230 Eucalyptus simulacra in Gawahon Eco Park

14.1.4 Beekeeping Map

Figure 231 Locations of Meliponiculture interest in Gawahon Eco Park, Victorias City, Negros Occ.

14.2 Ricky Sobesta's Meliponarry:

We met the beekeepers the previous night but those practising in Gawahon are Ricky Sobesta and Emmanuel Nisol. The map of their locations is appended (Fig. 39). The park boasts some *Eucalyptus simulacra* and endemic *Dipterocarpus* species, (for local names see Appendix)

Ricky has his Meliponary in his backyard and has collected some feral "Takpal" hives. He is also experimenting with eduction (Fig 232)

Figure 232 Ricky has his Meliponary in his backyard and has collected some feral "Takpal" hives. He is also experimenting with eduction

Figure 233 Takpal feral nest

14.3 Nisol Bee Farm

We were brought to an organic Farm within Eco Park. The owner Emmanuel Nisol was not back yet but we were allowed to browse around the Meliponary. They employ stingless bees to help in the pollination of their crops. We browsed the farm and quite obviously it was well planned and well maintained. Structures were well designed. At the end of the tour, we met the owner Emanuel Nsol at Freddie Lozada's place. His Meliponary was rather slipshod on bamboo racks. Unlike the other structures, this was like a temporary setup. The boxes were made of plywood and lacked the better construction of his other buildings.

Figure 234 Peculiar nest entrance of the "Takpal;" colony

Figure 235 We browsed the farm and obviously, it was well planned and well maintained. Structures were well designed. At the end of the tour, we met the owner Emanuel Nsol at Freddie Lozada's place.

Figure 236 His Meliponary was rather slip-shod on bamboo racks. Unlike the other structures, this was like a temporary setup.

14.3.1 Meliponiculture:

We did not open any hives for two reasons, one was because the owner was not around, and another was because I was worried the box might just fall apart. As it was the bees were already coming out of the seams and box joints. We were told by Freddie that the species were the layered brood type they called Takpal. Wish the owner was more concerned about the box construction.

14.4 Okinawa House

Figure 237 Beside the Okinawa house are a couple of traditional bamboo huts and a path that leads to Freddie Lozada's Meliponary

Victorias City established a sisterhood with Nanjo City, Okinawa, Japan in 2013, and with that, an NGO from Nanjo (NPO Lequoiwing foundation) had a partnership program finance by JICA with Victorias City on agri-eco tourism enhancement based on the Nanjo City model. Various sector representatives from Victorias and Nanjo had an exchange visit and tour of both cities to learn about Agri eco-tourism to be implemented in their place, so that area with Okinawan is dedicated as Victorias-Nanjo City Friendship Park, as a showcase centre of culture, art and a product of both cities in the near future.

The bamboo hut will serve as overnight accommodation for visitors, and more support facilities will be put up as part of the development plan.

Freddie Lozada's hives appear quite healthy and active on metal racks, a superior and sustainable setup for a Meliponary. He has an ant nest shell with a Takpal cohabiting in it. It was drizzling when we were there so we could not fully appreciate his setup.

14.5 Freddie Lozada's Meliponary

Figure 238 Freddie Lozada's hives appear quite healthy and active on metal racks. It was a more sustainable setup for a Meliponary

14.6. Fejumar A. Gasper, Capt Ramon Brgy, Silay City, Negros Occ.

14.6.1 **Getting there**: After striking a deal with Emanu el Nisol, he agreed to drive us in his 4WD. We were there in a couple of hours. Jumar's house was quite a drop from the road. It was naturally terraced. At the road level was a stall of sorts though not in operation then. The second level was his garden and Meliponary. The third level was his house with box hives hanging from the roof eaves. The fourth lowest level was the kitchen and dining area. That is where we met his father who had prepared lunch for us.

14.6.2: **Etymology**: The appellation Silay is derived from the name of a native tree that grew abundantly in the area. The kansilay is the city's official tree.

14.6.3 Meliponiculture:

Figure 239 A quaint garden with flower beds and a large square bed of Cosmos caudata.

A quaint garden with hives on stilts at the next level below the main road. The far end of the photo on the right is the entry down from the road. The right photo shows a large bed of *Cosmos caudata* at the base of his main feature tree.

After admiring his garden, and his hives on stilts (which by the way is good for ease of deterring predators like frogs, lizards and ants) he then obliged us by opening a few boxes while Emanuel, Romeo and myself looked on. I was curious to observe firsthand how the Northern Negros endemic bees developed in the suburb setting, unlike Gawahon Eco-Park which was more of a forest fringe setting. He was also into the eduction of a subterranean nest. I mentioned the hose was too long.

Figure 241 One of them came by and greeted me but I was not fast enough to get a sharp enough focus, but managed to pose at different angles and all I could make out was its genus (as if we did not know that) Tetragonula sp.

Figure 240 After admiring his garden, he obliged us by opening a few boxes while Emanuel, Romeo and myself looked on. The photo on the right shows he was also into the eduction of a subterranean nest. I mentioned the hose was too long.

Fig 4 Some of Jumar's box hives of the 'Takpal' kind the locals identify as a species quite similar to T. biroi's habit of congregating at the entrance of the box hives.

Figure 242 Some of the nest entrances were crusty looking and this reminded me of the T. cf valdezi that I have back home.

Figure 243 Hives at the house roof eaves at the lower 2nd level

Going down to his house at the lower level, we saw that Jumar hung more hives at the roof eaves, He had special apparatus to put up and bring down because they were quite high to reach.

14.6.4 Comparing brood arrangements. The left and middle show brood only at the corner implying the base has either some contamination or fungus. The brood arrangement of the new brood appears to have

Figure 244 Comparing brood arrangement, the left and middle show brood only at the corner implying the base has either some contamination or fungus. The brood arrangement of the new brood appears to have connectors and the pupa cocoons below are single cells yet clustered in different layers. On the right is the compact layered Takpal brood with an open brood chamber

Figure 245 The internal structure of one of his T. cf sapiens

connectors and the pupa cocoons below are single cells yet clustered at different levels. On the right is the compact layered Takpal brood with an open brood chamber.

Figure 247 The top three of this batch look strong and most likely a Tetragonula cf. laeviceps, Bottom left is another "Takpal". The bottom middle is clustered in layers and the bottom right is randomly clustered

Figure 246 Jumar showed some of his innovative trials while others may not be developing as expected

14.7 November Canieso-Yeo, Green Ville1 South, Sigma St., Bacolod City

14.7.1 **Etymology**: Bacólod (English: Bacolod), is derived from bakólod (Old Spelling: bacólod), the Old Hiligaynon (Old Ilonggo) (Old Spelling: Ylongo and Ilongo) word for a "hill, mound, rise, hillock, down, any small eminence or elevation", since the resettlement was founded on a stony, hilly area, now the barangay of Granada.

14.7.2 Hive Transfer: We were invited into her house for some refreshments and a small talk about the beekeeping industry in Negros. Soon after we proceeded to the main purpose of being there and which was for Jumar to help her transfer a colony from a deteriorating box to a new one that he brought. We went to her backyard to get it done.

November posed in her newly made bee hat and screen as she was keen to try it out and insisted to do the initial steps of the transfer. As it turns out her net was somehow penetrable by the bees and her face was swarmed by the attacking bees inside her bee hat net. There was quite a commotion while I retreated yet the bees were still after me, also because Ma'am November in search of relief thought the smoke from my cigarette might help calm them down. Unfortunately, the smoke was negligible to the bees and they even got into my hair and beard.

This only proves my advice to always make a proper box hive that is durable and can last many years, otherwise one has to continue doing such transfers to new boxes every time the previous one deteriorates. One should avoid using plywood. So, in any case, Ma'am November had to abandon the task at hand. Jumar put on the bee hat net and finished the job.

It was a moment to remember and fortunately was something we could laugh heartily about as Jumar was still having trouble getting the bees out of his hair and clothes. Outside the house, Romeo helped him build an open smoker by burning some leaves and plant debris while Jumar stood over the smoke. Soon when all was calm and cleared, we decided it was time to move on to Aboy's restaurant to meet its owner Nestor Evaristo.

Figure 248 Meeting Ma'am November for the first time in front of her house. Later Jumar prepared the box for transfer, November posed in her newly made bee hat and screen as she was keen to try it out and insisted to do the initial steps for transfer.

Figure 249 As it turns out her net was somehow penetrable by the bees and her face was swarmed by the attacking bees inside her bee hat net. Unfortunately, my cigarette smoke was negligible to the bees and they even got into my hair and beard. Some propolis was gathered by Ma'am November later on.

14.7.3 Addendum: Philippine Cinnamon and Philippine Stingless Bee Honey Tea

The Philippines is endowed with 19 endemic species of cinnamon, four of which I have seen in the Visayas.

1. Species A - Initial ID - *Cinnamomum anacardium*

Image 39 Species A - Initial ID - Cinnamomum anacardium

2. Species B - Initial ID - *Cinnamomum panayense*

Image 40 Species B - Initial ID - Cinnamomum panayense

3. Species C: Initial ID is *Cinnamomum iners*

Image 41 Species C: Initial ID is Cinnamomum iners

4. Species D: Initial ID - *Cinnamomum mindanaense*

Here's Plantsville Health's recipe for the tea:

1. Boil 300 ml water. Lower fire.
2. Drop 1-2 grams of Philippine cinnamon in chip form. Simmer for 5-8 minutes.
3. Tips - air dry only the bark (do not dry in full sun), and simmer on a low fire when brewing.
4. Add 1 tablespoon of stingless bee honey. Enjoy!

Image 42 Species D: Initial ID - Cinnamomum mindanaense

8.5 The flavour of the Philippine cinnamon is minty and warm, while the flavour of the Philippine stingless bee honey, the one I tasted at Talos Natural Farm and Bee Sanctuary(p. 158), is fruity. Together, the cinnamon honey tea tasted... phenomenal.

Both cinnamon and honey are not only great tasting foods, but they are medicinal, too. If one has a stomach ache (from diarrhoea, flatulence, indigestion, etc), drinking this tea makes the ache go away. When one is coughing or has a fever, this tea provides welcome relief.

I invite the readers to visit the Visayas for the cinnamon honey tea expedition. It's a unique experience.

Write-up and photos by:
November G. Canieso-Yeo
Plantsville Health, Bacolod City, Negros Occidental
+63 9175418462, +63939 9187020
novembercanieso@gmail.com

Figure 250 Cinamomou cassia

**Footnote: Interestingly an Australian stingless bee, *Austroplebeia cassiae* was named after a *Cinamomou cassia - Cinnamomum cassia*, called Chinese cassia or Chinese cinnamon, an evergreen tree originating in southern China, and widely cultivated there and elsewhere in southern and eastern Asia. (AHJ) (https://en.wikipedia.org/wiki/Cinnamomum_cassia)

14.8 Nestor Evaristo – Owner of Aboy's Restaurant

14.8.1 **Getting there**: Ma'am November together with Romeo and Jumar hoped on the back of Emanuel's 4WD and we headed for Aboy's Restaurant. I was told he had some bees in their courtyard. Sir Nestor Evaristo keeps bees in Don Salvador Benedicto (DSB) we lost daylight because sundown was fast approaching, so we opted to go to his restaurant. Jumar and Romeo volunteered to survey the DSB meliponary the next day on their own.

14.8.2 Meliponiculture: So, at the restaurant courtyard, we met Sir Nestor at his novelty rack of stingless beehives at the side corner of the restaurant courtyard. The vegetable plots and some plantings were there. The first thing he asked me was why his bees were in a bad state of health. I guess it was a test and he also wanted to know the purpose of my visit. I calmly viewed his hives and just smiled and greeted him cordially.

Figure 251 He recognized Sir Emanuel Nisol and they exchanged some pleasantries, I continued inspecting his hives and I presume all present waited for my answer

Figure 252 A friendship ensued, while we appreciated his "edible garden"

He recognized Sir Emanuel Nisol and they exchanged some pleasantries, I continued inspecting his hives and I presume all present waited for my answer. What I noticed was that all his boxes were built from half-inch plywood. I told him about the negative influence the plywood glue has on the bee's pheromones. He quickly got the idea and soon did not appear apprehensive anymore.

14.8.3 Aboy's Restaurant started in 1992 as a small eatery near the Burgos Public Market in Bacolod City. Initially, it catered mostly to travelling salesmen, cab drivers and office workers around the area. It offered home-cooked meals and fresh seafood that are abundant in the coastal towns of Negros. Soon word spread and as they say, the rest is history.

14.8.4 Talos Natural Farm in Don Salvador Benedicto town of Negros Occidental came into the picture 10 years ago when Nestor 'Aboy' Evaristo wanted to use naturally grown veggies for his restaurant. Don Salvador was the most logical choice having an elevation of more than 900 ft above sea level. It was in the course of his quest to create a natural environment for his farm that he decided to raise stingless bees and multiply them. As of today, Aboy keeps around 250 hives of "mungo mungo" species mostly of *laeviceps* and *sapiens*. Identification of other species is still underway.

14.8.5 **Don Salvador Benedicto – Etymology**: The town is named after the former Vice-Governor Salvador V. Benedicto (March 31, 1889–November 28, 1956) who played an important role in the setting up of a Revolutionary Government in Negros Island particularly in Barangay Igmaya-an, when the country was occupied by the Japanese during World War II.

14.8.6 Meliponiculture:

Image 43 Talos Natural Farm and Bee Sanctuary in Don Salvador Benedicta

14.9 BuroBuro Springs Vermi Farm

This farm is owned by Jose (Batchoy) and Pamela (Pam) Henares who signed into organic farming after they retired from the corporate world. It has been organic since 1997 and is an ATI Learning Site, Accredited Agritourism destination, and TESDA accredited Farm School Learning and Assessment Center. It has a rainforest preserve, fruit trees, vegetables and now stingless bees. They plan to incorporate strawberries with stingless bees in a net house.

14.9.1 **Getting there**: It is located at Barangay 5, Alangilan, Bacolod City and Talisay City, Negros Occidental.

14.9.2 **Etymology**: BuroBuro means bubbling water, which there is.

14.9.3 Interview Q & A

Q1. Where is the source of your bees? Forest fringe, village trees, hive hunters?

Ans 1: Some of the bees I bought from beekeepers and some from people who repaired their houses.

Q2. How long do you have been keeping bees? What prompted you?

Ans 2: I started with *Apis mellifera* some 20 years but I stopped when they swarmed.

Image 44 BuroBuro Springs Vermi Farm
This farm is owned by Jose (Batchoy) and Pamela (Pam) Henares

Q3. What design of hives are you using? Dimensions of the box?

Ans 3: My photos show what I use.

Q4. Where do you keep them? Residence garden yard, meliponary farm, or just racks in the backyard?

Ans 4: I keep them on the farm but I need to plant more flowers.

Q5. What sort of vegetation or foraging sources did you provide or place near wasteland shrubbery?

Ans 5: We have an organic farm and there is a lot of vegetation around. Thank you.

14.9.4 Meliponiculture:

Image 45 Some nest entrances found at the BuroBuro farm

14.9.5 Beekeeping Training

Stingless beekeeping training with organic farming students sponsored by TESDA and ATI at Buroburo Organic Farm -Talisay City.

Image 46 Stingless beekeeping training with Organic Farming students sponsored by TESDA and ATI at BuroBuro Organic Farm -Talisay City.by Fejumar Gasper and Freddie Lozada

14.10 Omoyon Bee Farm, Dumaguete, Negros Oriental

Omoyon Bee Farm is a backyard bee farm about eight hundred square meters inland area that has a mushroom growing area, a freshwater pond, mini piggery, fruit-bearing trees, and of course the honeybees and stingless bees.

It was 2010 when the father and son duo of Fred and Rico Omoyon started to keep stingless bees that colonized a number of their dried bamboo poles that were placed in a section of the residence. Stingless bees were not new to them as it was a daily sight for them in the early eighties at Fred's ancestral home which was all made of bamboo. They used to count colonies back then and the count would always reach above 60 colonies/entrances among the numerous bamboo cavities.

They kept the stingless bees ten years ago without knowing how to care for them. His son formally studied beekeeping at a state university and eventually ended up pursuing additional beekeeping short courses like honeybee queen rearing and honeybee queen breeding via artificial insemination at Harry H. Laidlaw Jr., Honeybee Research Facility, at the University of California, in Davis, USA.

Image 47 OBF Fig. 1.01

Image 48 OBF Fig. 1.02

Rico started MILEA Bee Farm in Batangas which is one of the country's known beekeeping entities to date. For stingless bees, Rico's love for them was nurtured even further after attending the first international stingless bee conference held in Malaysia. He got the chance to be toured around the forest reserve behind Genome Malaysia to see the different Malaysian stingless bee species by no less than Abu Hassan Jalil, and that started a deeper stingless beekeeping venture for Rico.

In Dumaguete City, they use coconut shells from matured coconuts as a hive for stingless bees. It is a completely different method from that practised in the Bicol Region.

Asked about his choice of coconut shells for stingless bee hives, Rico explained that he prefers the material as it is all-natural and that bees have been known to colonize the same. Besides, it is also naturally available in their area as a coconut-growing province. He believes that dried and cleaned coconut shells do not emit any insect-repellent scents making it an ideal hive material candidate. Regarding its durability, Rico admits that, unlike other seasoned stingless beekeepers who design the thickness of hives to last for ten years, his coconut shells may only last him four years but then, his materials are always available and he can readily replace a damaged hive, Rico explained. Coconut shells are not that thick but solid enough that termites would let them pass, Rico continues. Please note that Rico does not remove the husk of his coconut shells as he claims that it helps to insulate the bees and the coconut shell as a whole. He would usually find fungus, if and when, the coconut-shell husks are removed, he says.

Image 50 OBF Fig. 1.03

Image 49 OBF Fig. 1.04

Other issues and concerns about using coconut shells as a hive for stingless bees are discussed below, with photos.

Omoyon Bee Farm strives to grow its colony numbers without harming the present biodiversity of a location. They avoid extracting colonies from the wild, but they are achieving their aim by installing coconut-shell bait hives in locations where stingless bees are spotted.

Fig, 1.01 and Fig, 1.02 illustrate how Omoyon Bee Farm coconut-shell bait hives would look. An entire coconut shell is mounted on a base (that is about half of the coconut shell used as a hive). We know for a fact that empty coconut shells are light thus, stability against the wind may be in question but that was already addressed, as the base of the bait hive is full of concrete for weight. It can withstand strong winds because coconut shells by themselves are naturally aerodynamic, being round. The only problem is that cement is known to retain daytime heat until night and so to address that, Rico would always install his coconut-shell hives under the shade – away from sun and rain.

Image 51 OBF Fig. 1.05

Rico laments that the average weight of an empty coconut-shell bait hive would be around 1.2 kg, as illustrated in Fig, 1.01.

Image 52 OBF Fig. 1.06

Rico always goes around the province looking for favourable locations. When an area is chosen for the deployment of coconut-shell bait hives, he then schedules a beekeeping orientation with the landowner to discuss the benefits of having pollinators around.

After securing the landowner's green light, Rico's team would then strategically install at least ten bait hives in each location. A strategic location for them would mean that they first evaluate the wind direction and that bait hives will be situated a few meters in front of the colony in its natural habitat, along the wind-directional path. Each stingless bee bait hive has a sticker underneath the base that contains the following information:

1. Location of the installation
2. Date of installation
3. Weight (or the empty stingless bait hive)

Image 53 OBF Fig. 1.07

Once the stingless bait hive is colonized, Rico rewards the landowner with One Hundred Pesos for each colonized bait hive as part of the agreement that will benefit both the parties concerned. The new colony in the bait hive will then be monitored and evaluated for strength by way of inspection and the weight of the stingless bee bait hive. When the colony is strong enough, it will then be sealed at night for transport to the Omoyon Bee Farm home base. Sealing the bees inside is as easy as One, Two, or Three, because the Omoyon Bee Farm stingless bee bait hive (for Mongo-mongo) features an easy-to-close entrance as it is installed with the nozzle of the soda bottle (shown in Fig, 1.03). When the bees are sealed inside, stress is lessened as aeration is amply provided by the numerous air passages at the soda nozzle, illustrated in Fig, 1.04.

The reason for bringing the inhabited bait hive to the Omoyon Bee Farm home base is for the installation of a second level (another coconut shell on top of the bait hive) to give the bees additional space. The new coconut "super" will be for honey (and sometimes pollen) only. A stingless bee queen excluder will also be fabricated, to restrict the queen to the brood area, before the installation of the honey super. Fig, 1.05 illustrates what this would look like.

Image 54 OBF *Fig. 1.08*

To harvest, you only need to open the topmost half of the shell to expose the honey pots to perform your preferred harvesting technique either by Suction, Prick and Pour, or the Squeeze and Strain harvesting method.

It may be asked why Rico only has an entire coconut shell (both halves as shown in Fig, 1.05) for a honey chamber as compared to an average of eight halves practised in Bicol. He answers that it is because *T. laeviceps* is the most common species in Dumaguete City and Oriental Negros in general. A single coconut shell may not be enough for a honey chamber if we think about the photos circulating on social media showing *T. laeviceps* or mongo-mongo stingless bees occupying and filling two bamboo nodes but then Rico explains that those photos were taken from a feral colony that may have not been opened and harvested for the past three years. And so, he thinks that an entire coconut shell would suffice as a honey chamber especially since he intends to harvest yearly. Also, he adds that he can perform an early inspection (end of February) than his regular harvest period in May. By doing the same, he can evaluate the need to either fabricate a new queen excluder before adding a third layer (honey chamber) or just continue the usage of the second layer (honey chamber) until the harvest period.

Image 55 OBF *Fig. 1.09*

The hives discussed earlier were Rico's coconut shell design for the mongo-mongo stingless bees but he has also experimented on a hive designed for "Takpal" and *T. biroi* as shown in Fig. 1.06 and Fig. 1.07, respectively.

Fig, 1.07 shows the use of three coconut shells with the brood/entrance installed in the second level (centre) and its honey chambers are situated in the first and third chambers. This hive design features two queen excluders (the topmost portion of the first and second chambers). Curved aluminium bars are installed to lift the brood and to leave ample space at the bottom for the workers going in and out of the lower honey chamber, as shown in Fig, 1.08. Rico added that the efficiency of this hive design is yet to be tested; this will take at least two consecutive honey seasons.

Image 56 OBF Fig. 1.10

Moving around Omoyon Bee Farm, one would notice how Rico loves to experiment on coconut-shell hives for the "Visayan" stingless bees. Now that his location is already frequented by visitors as a form of farm tourism, he also developed a coconut-shell hive design for the visitors' delight, shown in Fig, 1.09. It looks fancy but is still functional as the two chambers above also have queen excluders as shown in Fig, 1.10.

Omoyon Bee Farm is gearing towards the improvement of their coconut-shell hives while experimenting on other hive types as it welcomes the growing number of tourists appreciating "the bees of the future", the stingless bees. If interested, it is recommended that a visitor first book an appointment before BEEing with them at Sto. Niño Road, Banilad, Dumaguete City, Negros Oriental.

Coconut tree trunk Bait Hive; A personal account by Rico Pietro

The 1987 study mentioned the house of Anastacio Umbac in Dumaguete with at least 84 colonies. The same research observed '*mungo-mungo*' i.e., *Tetragonula* spp. in coconut trees in Valencia ... and that is where I usually park my bait hives. Usually, almost all coconut trees have at least a colony, but not today. I came early to install my bait hives. Sad to see the damage done to the bees.

Image 57 Bait hives - coconut tree trunks

Image 58 Coconut Trees in Valencia, Negros Oriental

Corn grit foraging

By the way, in the upland municipalities here, you can see a large volume of stingless bees in the market specifically in rice stores but they are gathering the tiny corn grits ... guess they are taking it as pollen substitute or the sugar content.

Yes, we feed them corn grit as dry feed. This solves ant and fly problems compared to giving liquid or mush feed.

Image 59 Corn grit at municipality markets

14.10.1 Dumaguete Stingless Beekeepers Group:

Rico Pietro Omoyon
Sto. Niño Road, Banilad
Dumaguete City
312 colonies

Tata A. Delgado II
Tigabawan, Lazi, Siquijor
6 colonies

Lauro Gonzales
San Jose, Banilad
Dumaguete City, Negros Oriental
8 colonies

Helmar A. Mariño
Sto. Niño Road, Banilad, Dumaguete City, Negros Oriental
4 colonies

Celso Bejec
Caratagan, Dalaguete, Cebu
7 colonies

Don Eugene T. Gara
Dauin, Negros Oriental
7 colonies

Jared Poblador
Good Infinity Farms
Lot 2001, PLS-759, Tugawe, Dauin, Negros Oriental
48 colonies

Dr. Patrick Chua
Batinguel, Dumaguete City
9 colonies

Chapter 15 – Panay Island

15.1 Iloilo Province, Panay

15.1.1 **Getting there**: We planned to get the first ferry out to Iloilo from Bacolod but getting there before 6 am proved arduous so we opted for the 9 am ferry after breakfast at least. We arrived at Iloilo port around noon.

15.1.2 **Etymology**: Irong-Irong (or Ilong-Ilong) appears in the Maragtas legend of the coming of ten Bornean datus (i.e., chieftains) to the island of Panay. The datus bartered with a local Ati chieftain for the plains and valleys of the island, offering gold in return. One datu, Paiburong, was given the territory of Irong-Irong, which is now the province of Iloilo in the Philippines. According to Spanish written accounts, the inhabitants of Panay Island hailed from Borneo, Sulawesi, and Sumatra. Panay might be named after the kingdom of Pannai which is located in Sumatra since i and y are interchangeable in Spanish. Proof for this is corroborated by linguistic evidence such as the Talang Tuo inscription where many Old Malayan words during Srivijayan survived in the modern Ilongo Language.

Hispanicized corruption of *irong-irong*, Hiligaynon for "nose-like ", referring to the shape of the delta formed by what is now called the Iloilo and Salog Rivers on which the settlement of the same name thrived. The name originally only applied to the town (now city) of Iloilo (rendered in Spanish orthography)

15.2 Perla Cabalfin, Talacuan, Leon, Iloilo

15.2.1 **Etymology**: Leon is the first name of Greek origin-the Greek λέων (léon; leōn), meaning "lion", has spawned the Latin "Leo", French "lion", Irish "Leon" and Spanish "León". ... A similar Greek name to Leon is Leonidas, meaning "son of a lion", with Leonidas I, king of Sparta, being perhaps the most famous bearer of that name.

The municipality of Leon was formerly known as Camando which was founded in the year 1730. In 1866 the town was transferred to Sitio Capan, a vast rice region and after two years it was given the name Leon. There were two versions of how the town was named: first, derived from the first name of a prominent member of principalia, Captain Leon and second, the name Leon was given in honour of the City of Leon in Spain, the place where the Parsh Priest of the town came from.

15.2.2 **Meliponines**: One colony was found in a Bamboo pole and had a double entrance. Another was on the window edge of her house. For the bamboo hive, they can transfer it into a box and for the one on the window ledge, they would have to try the eduction method. In Both cases, NAP will assist and guide them since they were quite close in proximity to NAP's residence.

Figure 253 one colony was found in a Bamboo pole and had a double entrance. Another was on the window ledge of her house

15.3 Herman Caololan in Awis, Leon, Iloilo

15.3.1 Meliponiculture: Herman found some feral colonies on old tree trunks and a couple in bamboo internodes and NAP has guided him to keep one in a box. As I was there incidentally, I was able to help him transfer a bamboo colony.

Figure 254 Inspecting the box hive and the brood looks good and developing well. T cf. sapiens

Figure 256 Herman's beautiful bamboo hut with a shed outside his hut

Figure 255 Assortment of feral hives

15.5.3 Bamboo Hive transfer

Figure 257 70 yr. old ancestral home with woven bamboo wall panels with numerous SB nests sporadically; in Awis, Leon, Iloilo

15.4 Lucia Gadian of Brgy. Awis, Leon

Figure 258 There were nests in almost every internode of the bamboo wall panel, 109 in total

15.4.1 Meliponines:

The house was a typical Tausug traditional Filipino bamboo architecture. There were stingless bee nests in almost every internode of the bamboo wall panel, 109 in total, personally counted by NAP.

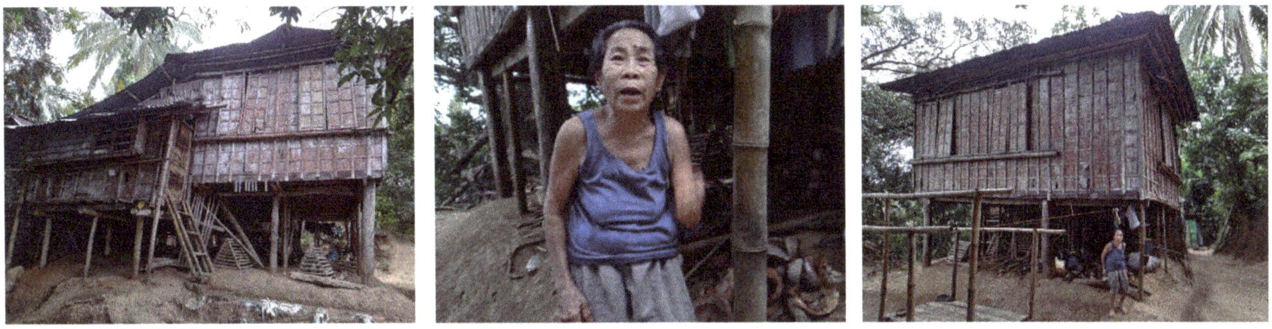

Figure 261 Lucia Gadian at 82 yr. old in front of her 70 yr. old Tausug Architecture House

Figure 259 We travelled on motorcycles and stopped for a breather to enjoy the panorama. With Herman at the Century old Mango tree

Figure 260 Rudolf joined in collecting SB nests from the wall panels

15.4.3 On Feb !0th, NAP did a follow-up visit to Awis together with Marc Lyndon (Rudolf Rnr of Guimaras.)

The Video link is appended below-

https://www.facebook.com/nora.pena.963/videos/10218466529767788/?t=0

https://www.facebook.com/nora.pena.963/videos/10218466673731387/?t=2

Figure 262 Nearby, Herman checked out a nest entrance in the ground. Rudolf was there and at his insistence, Herman started to dig it out

Figure 263 The honey was packaged and sealed with the help of Rudolf. The involucrum was collected for drying and processing

Figure 265 After digging up two nests, the results are some pollen pots, honey pots, and brood from one of the nests

Figure 264 We collected the involucrum and salvaged another strong mature brood and packaged the honey collected

Figure 266 Harvested, Bottled and labelled as Visayas Stingless Bee Honey

15.5 John Cagalitan of Sitio Balud, Población, Leon, Iloilo

15.5.1 Meliponines:

Figure 267 John rears pigs and runs a sawmill as well. The bees were found in the bamboo rafters of his pigsty

John rears pigs and runs a sawmill as well. The bees were found in the bamboo rafters of his pigsty. At his lumber stockyard, we found some old *Gmelina* lumber which I told them would be excellent for box hive fabrication. Just needs shaving and ripping. The insides can be flamed and propolis extract treated.

Figure 268 old Gmelina lumber would be excellent to make thick box hives as they are already well seasoned. No warping issues

15.6 Alan Cabalfin of Camillas St., Leon, Iloilo

15.6.1 **Getting there**: Camillas Street is not far away and is perpendicular to Tabiana street (NAP's residence). We took a tricycle to get there.

15.6.2 **Pottery**: Alan Cabalfin is an accomplished potter who just came back from the USA. He has his kiln and dabbles in terracotta and ceramics. I was interested in getting him to work on terracotta for hive containment and also small honey jars.

Figure 269 Top left: Alan's house is a testament to his artistic touch. Bottom left; Alan's kiln in his workshop and some of his ceramic work

15.6.3 Meliponines: Alan has one hive which to me looks like a typical NAP design although NAP now has changed to solid wood compared to plywood.

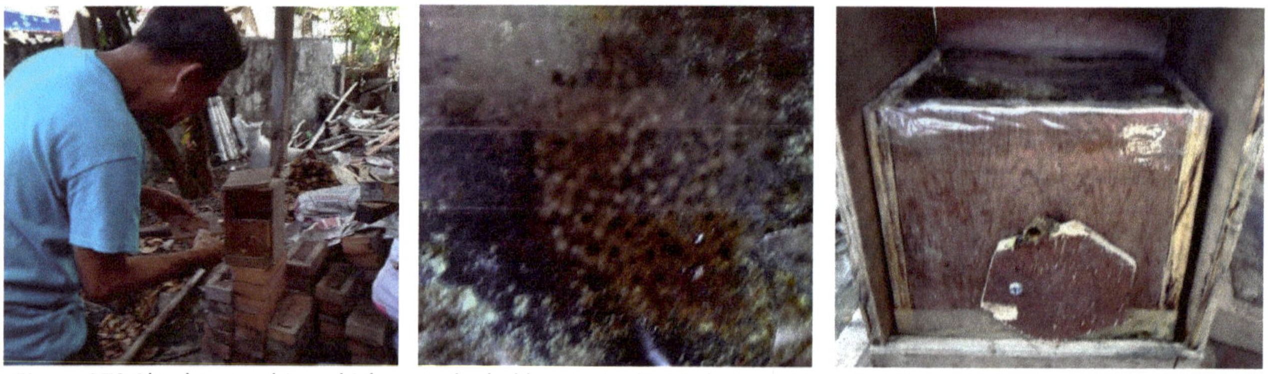

Figure 270 Alan has one hive which to me looks like a typical NAP design although NAP now has changed to solid wood compared to plywood.

15.7 ISAT-U LNCA Campus, Leon, Iloilo

15.7.1 Arriving at the Campus:

Figure 271 Arriving at the ISAT campus and meeting some of the lecturers and staff, then meeting with the administrator

Had the opportunity to meet with Mdm. Eugenita C. Capaciete, Campus Administrator of the Institute of Science and Technology University and Department of Agriculture Personnel from the Municipality of Leon were also there.

We requested permission to look at some lobations on the campus grounds to view some stingless bee nests. Some classroom doors had stingless bee nests. On the grounds, we also located a couple of trees that had SB nests.

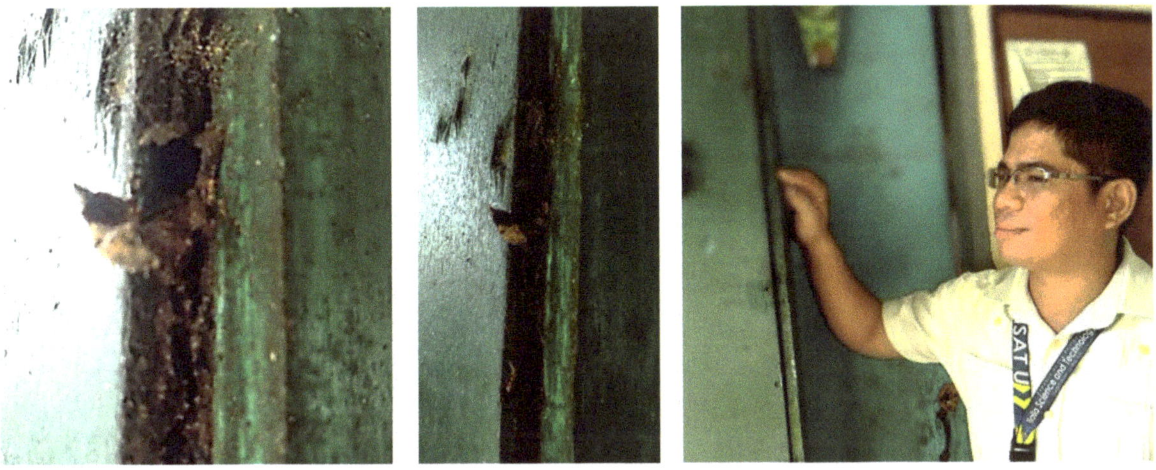

Figure 272 The door of Sir Apolonia Cabangal Jr.'s classroom

15.8 Bucari Pine Forest, Sitio Tabionan, Iloilo

15.8.1 **Getting there**: From ISAT we went to Sencia Clarito, Department Head, Dept. of Agriculture, Municipality of Leon. From there we were driven to the Municipality Nursery in Bucari.

We arrived at the Bucari Pine Forest after 40 min. and it was raining so we took refuge at the Pine Forest Campsite. After the rain stopped, we had to walk up to the Municipal Nursery about 500 m away. It was quite an exhausting climb for me.

15.8.2 **Etymology**: Bucari in Filipino means "Go ahead". It is also known as "Iloilo's Little Baguio" because of its cool climate. It has a campsite where visitors could pitch a tent and spend the night in the woods.

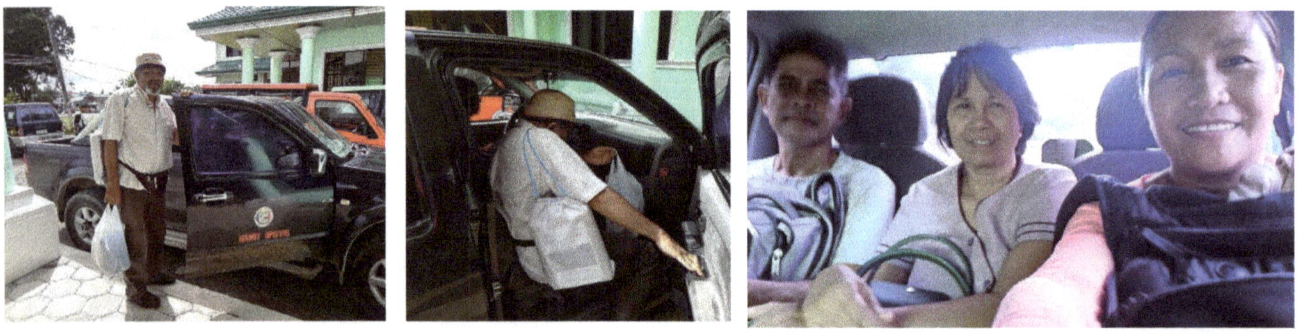

Figure 274 At the Dept. of Agriculture, Leon Municipal; GAMIT OPISYAL written on the car door means "FOR OFFICIAL USE" From there we headed for Bucari with Alan Cabalfin, Sencia Clarita (Department Head, Dept. of Agriculture, Municipality of Leon) and NAP

Figure 273 Bucari Pine Forest

15.8.3 Leon Municipal Nursery: From the Pine Forest we climbed up in the drizzle. I was more worried about slipping from the wet ground than worrying about catching my breath. It was quite daunting in a sense.

Figure 275 From the Pine Forest we climbed up in the drizzle, I was more worried about slipping from the wet ground rather than worrying about catching my breath. It was quite daunting in that sense.

15.8.4 Meliponines: Their Nursery staff had a feral SB hive in a Bamboo pole and one end had a Wasp nest, so they did not know how to transfer the SB colony. NAP brought one of her boxes to be used as a new hive. The following photos show the sequence.

Figure 276 After covering the Wasp nest end, I sawed the diameter of the bamboo in half before the node and pried it open to reveal a full healthy colony

Figure 277 With the wasp nest end covered with a plastic bag, I start to pick out the brood to transfer into the box NAP provided

Figure 278 I proceed to transfer the brood into the box with Alan's help and left out the honey pots for them to enjoy

15.9 Antonio Capirayan of Brgy Tacuyong Sur, Leon, Iloilo

15.9.1 Bamboo House:

Figure 279 60 yr. old Traditional Bamboo with the Tausug Influence. Bees simply love the internodes of the Bamboo wall panel frames

15.10 Esperidion Calaor ng Brgy. Gumboc, Leon, Iloilo

15.10.1 Bamboo House:

Figure 280 Another Traditional Bamboo house has 24 colonies

15.11 Ivan Paul Pauchano, of Barangay Cabanga-an, Badiangan, Iloilo

15.11.1 **Getting There**: From Tabiana street it takes 1 hour 45 min. via Iloilo Airport Road

15.11.2 **Eduction**:

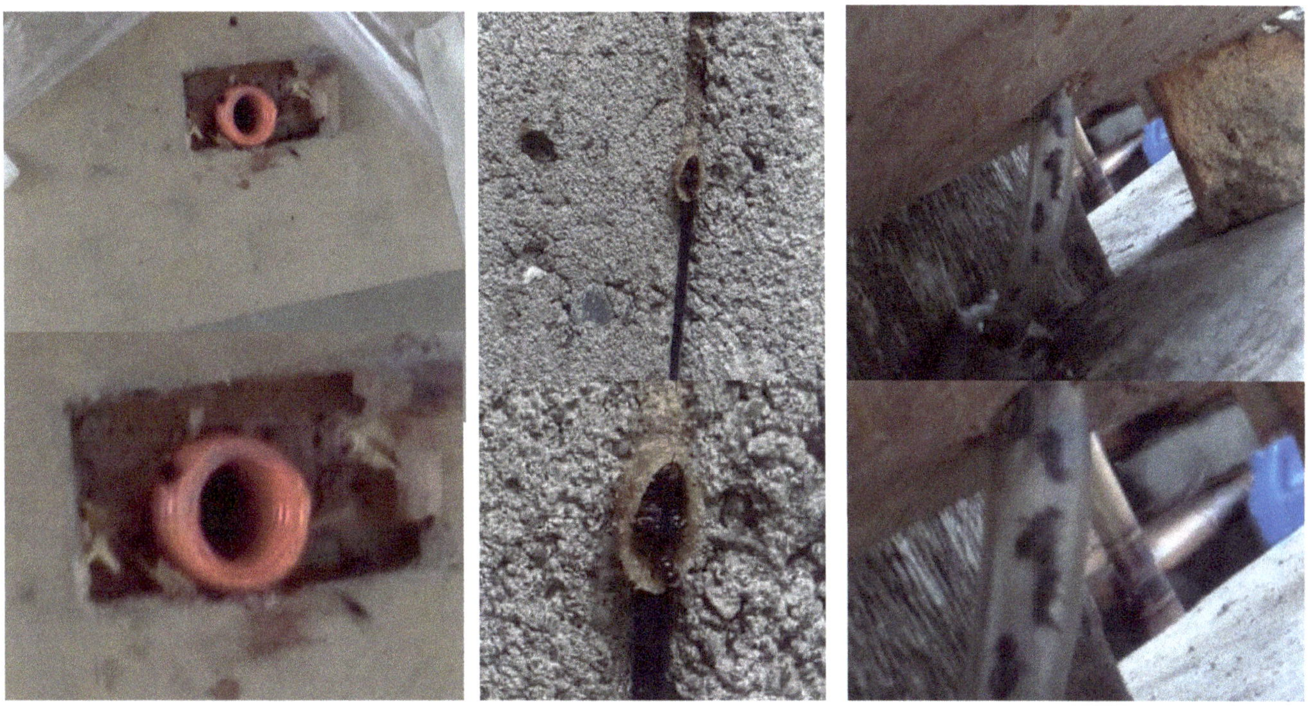

Image 60 Eduction trials on nest in the cement wall.

Ivan Paul Pauchano is a beekeeper who told us about finding some SB nests in the crack of the walls of his home in Barangay Cabanga-an, Badiangan, Iloilo. The images are his attempt at the eduction of those colonies.

15.12 Nora Alanza Peña (NAP's) Residence, Tabiana Street, Población, Leon, Iloilo

15.12.1 **Getting There**: Going to the NAP meliponary is very easy. If you are in the town centre, just approach any tricycle to bring you to the residence of NAP.

15.12.2: Meliponary:

Figure 281 Helping out transfer from the previous plywood box to a solid wood box. Alan Cabalfin and her neighbour Pastor helping out

Figure 282 Found a spider predator in a sunflower; Patola plot; lumber supply for making box hives

Figure 283 Alan helping to execute eduction of a nest in her bamboo rafter, examples of her solid wood boxes at 1.75 inch thickness

NAP is a remarkable Meliponiculturist with exceptional building skills. She makes her boxes using simple tools. The power tools she has are an electric portable drill and an angle grinder. She manages a vegetable garden with some sunflowers.

15.13 Meredith Arillo, Sitio Bulad, Población, Leon, Iloilo

A keen enthusiast of stingless bees, she clings on to her precious colony in the bamboo piece. I visited my best friend Meredith Arillo at Sitio Bulad, Población, Leon and she has shown me her first stingless bee colony. She is hoping to have more in the coming days.

Figure 284 The one precious feral bamboo colony in her possession

Chapter 16 – Samar Island

16.1 April Fatima Dira Villanueva, Brgy. San Joaquin, Calbayog, Samar

16..1.1 **Getting There**: It takes 4.5 hr. to get to Calbayog from Sorsogon City to Calbayog City by road and ferry although both cities have airports and it is a short one-hour flight. They also run an HB farm called "Apiaryl Honeybee Farm" and conduct Honey Beekeeping courses occasionally.

16.1.2. **Etymology**: 1) Taboc is the place where the name Calbayog began. Taboc, so legend says, once abounded with "Bayog" trees. They were cut down and burned for fuel in making "Cal" (lime) out of seashells and corals. From these two things, the Spaniards called the place "Calbayog".

The name "Samar" was derived from the local language samad, meaning "wound" or "cut", aptly describing the rough physical features of the island, rugged and deeply dissected by streams.

Image 61 Boxed T. biroi from Sorsogon, one coconut shell hive covered with a coconut spadix

16.1.3 Meliponiculture: They have *T. biroi* from Sorsogon but also have five feral colonies of mungo2 from Calbayog, Samar. They have been keeping SB for more than three years. The *T. biroi* is more aggressive, compared to the Samar Bees. The hive designs utilize a box and a bao hive (Fig. 147).

Image 62 Mungo mungo feral colonies occur in steel pipes (Top left photo: five iron pipes attempt eduction on each nest.), Bottom middle and right photos respectively: rock cavities and coconut tree trunks

16.2 Ruel Benitez Delantar, Maypangdan, Borongan City, Eastern Samar.

16.2.1 **Getting There**: By flight one can approach Calbayog City airport in Samar or Tacloban City, Leyte. Either way, it takes about 4 hours

16.2.2 **Etymology**: Pronounced bo-róng-gan, the name Borongan was taken from the local word "borong", which in the Waray-Waray language means "fog". The mountainous terrains surrounding Borongan are covered by a heavy veil of fog which can usually be seen during the cold and rainy seasons and in the early hours of the morning

16.2.3 Meliponiculture:

1. The Stingless bees were first discovered nesting inside an inverted stalk of a giant tree fern used to grow orchids. The bees nest in the rock wall and even in posts at home. This is the giant fern. Inside of these is a hollow where bees are nesting.

Image 63 Fern tree Cyathea sp,

2. Keeping bees was not my hobby until I was invited to a training on *Apis mellifera* at Benguet State University in La Trinidad. We had training for a week sometime in 2017. There I liked the stingless bees which were in an additional lecture that included hands-on keeping. Then I realized what we have at home.

3. I've tried box designs and tried some but now favour two pieces that measure 18 cm x 18 cm and an 8 cm heigh box that can be easily made for splitting purposes, and 2.5 cm for the base. Hives also have a clear cover for viewing.

Image 64 Box hives beside the house

4. I keep the bees at the side of our home and practice splitting to multiply them because I don't like to disturb SBs nesting on rocks and trees.

5. The type of vegetation here in our place is so diverse. Mostly the seasides are dominated by mangrove forests. Coconuts dominate in the upland. And the rest is forest. Sometimes the bees forage in bananas, acacias, aratilis and flowering grasses.

Ruel Benitez Delantar is currently an instructor at Eastern Samar State University. He is an Agricultural and Biosystems Engineer. He believes *T. biroi* cannot do well in his locality. They can't survive very high humidity. Almost 80% RH in our place. Moulds are fatal for them.

Image 65 Successful swarm trap and advancing brood in a box hive

Chapter 17 – Guimaras Island

17.1 Guimaras – Marc Lyndon Genovana, Jordan

17.1.1 **Getting there**: Guimaras is a small island formerly a sub-province of Iloilo. A short ferry ride from the port of Iloilo. We met Marc Lyndon or Rudolf as he prefers at Jordan port. we then took a Tricycle with Rudolf and headed for his Barangay Santa Teresa.

14.1.2 **Etymology:** Cebuano derivative is From Latin Iordanes, from Ancient Greek Ἰορδάνης (Iordánēs), from Biblical Hebrew יַרְדֵּן (yardén, “Jordan (river)”). https://en.wiktionary.org/wiki/Jordan#Etymology_2

The etymology of Guimaras, in turn, was that Guimaras was formerly called Himal-us from the word “paghimud-us”, meaning a struggle for survival. In the early part of the Spanish time, the name Himal-us was changed to Guimaras.

Figure 286 OTW to Santa Teresa from Jordan Port

Figure 285 Ferry trip from Iloilo to Jordan port, Guimaras

17.1.3 Meliponiculture: With his engineering background Rudolf engineers some well-designed hives in shoebox sizes, finished in whitewash. His bees are almost all *T. laeviceps* and a couple of *T. fuscobalteata* hives.

Figure 287 T. laeviceps appears well served in these shoebox-size wooden box hives. However, the spacing I thought was too close. There would be battling problems during the mating season and if adjacent hives were to swarm at the same time, there may be drifting and the bees would get confused

T. laeviceps appear well served in these shoebox-size wooden box hives. However, the spacing I thought was too close. There would be battling problems during the mating season and if adjacent hives were to swarm at the same time, there may be drifting, and the bees would get confused. I mentioned to him that it may be suitable to place them at least one swarm spere apart. To avoid drifting, having hives conspicuously numbered and coloured would be advantageous and less stressful to bees.

Figure 288 Healthy brood although a bit of fungus was revealed, in the mid-bottom photo

Figure 289 Healthy nest entrances

17.1.4 Vegetation provision:

Figure 290 Some of the flowering vegetation in his backyard

List of Figures in Part 3

PART 4 - BICOL REGION V

Foreword for Part 4

Today, the role of stingless bees and other non- *Apis* species in biodiversity maintenance through pollinators of wild and cultivated crops is highly recognized. Being native to Asia, stingless bees are fit pollinators of local plant species. In the Philippines, stingless bees are used in large-scale pollination of mango and other orchard crops. Moreover, raising them is sustainable because they are resistant to pests and diseases and have an ample genetic pool coming from the wild populations. While there are available scientific publications on the diversity of stingless bee species and their role in pollination, the publication of this handbook on Meliponiculture for Indo-Malaysian and Asean Regions is very timely. We need a sourcebook that will guide prospective beekeepers and practitioners in the management of these species. The practical aspects of meliponiculture shared by beekeepers enhance the value of this handbook.

I commend the author for his sustained interest in meliponines and advocacy to promote the protection of native bee species. The information from this handbook will fill the gaps in our knowledge of meliponines. May this inspire the younger generation to conserve bees and appreciate their value to agriculture and the environment.

DR. CLEOFAS R. CERVANCIA

Professor Emeritus and President APIMONDIA Regional Commission for Asia

Institute of Biological Sciences

College of Arts and Sciences

University of Philippines Los Baños College, Laguna

Acknowledgements

The Authors express gratitude to all the literary contributors in this compendium. We also acknowledge and appreciate the valuable guidance from senior advisors and friends namely, Dr. David W. Roubik, Prof. Deborah R. Smith, Prof. Cleofas Cervancia, Prof. Maria Dulce Mostoles, Alejandro C. Fajardo Jr. and Dr. Ador Picaza.

Preface to Part 4

These passages are excerpts from 2 volumes of "The Handbook of Meliponiculture". They are reproduced here to complete the compilation of our tours in the Philippines. They have been placed in two sections; the first section is excerpted from Volume 1 of the Handbook on the Philippines that I toured in 2016. The second section is excerpted from the Handbook Volume 2 of the tours in 2017.

The original aim was to study and compare the various methods of Meliponiculture in the ASEAN region. The articles herein start with a depiction of Meliponiculture in the Philippines up to 2016. From then to date, much has changed and developed; the number of beekeepers and bee farms concentrating on stingless bees has expanded impressively.

In section 2 we will see many more locations being explored. I visited Los Baños in Laguna Province where The University of the Philippines, Los Baños (UPLB) Bee Program exists, before proceeding to Naga city to visit the activities of Central Bicol State University of Agriculture (CBSUA) in Pili, Camarines Sur, Bicol. There I had the opportunity to interact with the Beekeepers Camarines Norte Association - Region V Bicol Philippines. Touring Bicol I ended up in Guinobatan, Albay and Sorsogon.

Part 5 looks at Meliponiculture in Mindanao with literature and images contributed by Ace Lonzaga and some studies on *Tetragonula biroi* flight patterns.

From Bicol, I visited Tagum City, Davao Region and with the help of some Beekeepers from Panabo, I was brought to visit Davao Oriental and then Davao Norte. The travels conclude with Part 5 with charts on the latest Meliponiculture activities in the Bicol and Davao regions of the Philippines.

The Bicol Region (Central Bicolano: Rehiyon nin Bikol/Kabikolan/Bicolandia; Rinconada Bicol: Rehiyon ka Bikol; Tagalog: Kabikulan), also known simply as Bicol and historically known as Ibalong before Spanish colonization, is a region of the Philippines, designated as Region V. Bicol comprises six provinces, four on the Bicol Peninsula mainland (the southeastern end of Luzon) – Albay, Camarines Norte, Camarines Sur, and Sorsogon – and the offshore island provinces of Catanduanes and Masbate.

Catanduanes

Hispanicized and pluralized form of katanduan, Bikol for "place abundant with tando trees", referring to the abundance of these trees on the island.

Masbate

Some early accounts record the name of the province's main island as Masbat. This may be based on masibát, a native word for “abundant with lances”, perhaps about the island's well-armed inhabitants; or on masabat, Bikol for “to meet along the way”, alluding to the strategic position of the town (now city) that bears the name, as well as the island named after it, within old Philippine maritime trade routes.

Other early accounts record the name of the island as Masbad, which may be based on masibad. In Bikol, this word means “to pass through from end to end”, thereby alluding to the island's elongated shape; in Waray, this word means “devourer”, alluding to the island's once crocodile-infested rivers.

Camarines (Norte and Sur)

Plural form of the Spanish word camarín, the term used by the Spaniards to refer to the storage sheds (kamalig) that were abundant in the fertile and densely populated Bicol River plain in what is now Naga City and central Camarines Sur.

Albay

The shortened form of the phrase al baybay, composed of a Spanish preposition and Bikol root word, meaning “by the shore”, refers to the coastal settlement of Sawangan, now the port district of Legazpi City. At the time it was shortened to Albay, and the name was applied to the province over which the town of Albay (now Legazpi City) served as the capital.

Sorsogon

Hispanicized form of sogsogon, Bikol adjective meaning “wadeable”, derived from the root word sogsog, which means “to wade” or “to ford”; a wadeable river, for example, is known as salog na sogsogon. The toponym originally referred to the settlement on the banks of the Salog River, which became the capital of the province upon its establishment in 1767.

The regional centre and largest city are Legazpi City. The region is bounded by Lamon Bay to the north, the Philippine Sea to the east, and the Sibuyan Sea and Ragay Gulf to the west. The northernmost provinces, Camarines Norte and Camarines Sur are bordered to the west by the province of Quezon.

Meliponiculture Distribution in Bicol Region V

Chart 7 Meliponiculture Distribution in Bicol Region V

List of Meliponiculture activities in Bicol Region V

1. Lope Buboy Soriao Jr., Labo Camarines Norte
2. Ro Na Lyn, Baud, Camarines Norte
3. Sherwin Baja, Naga City, Cam Sur, Bicol
4. CBSUA Meliponari, Naga City, Cam Sur, Bicol
5. Gregorio Morota Jr., Tanfarora, Guinobatan, Albay
6. Ronald Garcia, Ligao City, Albay
7. Cristina Imperial, Hindi, Bacacay, Albay
8. Henry Rafael, Ligao City, Albay
9. Johnny Bobadilla, Sorsogon City, Sorsogon
10. Leo Grajo, Grajos Bee Farm, Casiguran, Sorsogon
11. Edwin Banares, Sorsogon City
12. Jro's Sonnyandsonloft, Bulusan, Sorsogon
13. Luz Gamba, Balay Buhay sa Uma, Bulusan

Chapter 18 - Central Bicol

18.0 Central Bicol State University CBSUA, Pili

I was brought to the Central Bicol State University in Pili.

Figure 291CBSUA Meliponary across the road from its Main Campus grounds.

Figure 292With CBSUA Meliponary Beekeeper Mr. Bert and Mr. Glen under a trellis of passion fruit creepers. Delivered a Meliponine conservation presentation at CBSUA with local and regional beekeepers & entomology students in their Internationalization of Beekeeping Advocacy in Bicol Region Program

18.1 Meliponiculture Initiatives in Bicol Region

Meliponiculture Initiatives Of The Central Bicol State University Of Agriculture, Bicol Region, Philippines by Dr. Maria Dulce Jovillano-Mostoles

Professor, Central Bicol State University of Agriculture, Pili, Camarines Sur

ABSTRACT

The Central Bicol State University of Agriculture (CBSUA) formerly the Camarines Sur State Agricultural College (CSSAC), as its name suggests, is an institution dedicated to research and higher education in agriculture. It runs the Regional Apiculture Center (RAC) for Bicol and it is located in the municipality of Pili, Province of Camarines Sur, in the southeast of Luzon. Its sole mandate is to promote beekeeping technology emphasizing keeping stingless bees, tagged as Meliponiculture.

The RAC for Bicol is manned by a multidisciplinary team of dedicated faculty members involved in research dealing with the biology and culture of the stingless bee species, *Tetragonula biroi*. Together with smallholders of local communities, a technological package was developed to keep these bees for their economic exploration, using halved coconut shells as the bee's nesting cavity, with innovations introduced. Also, learner-friendly methods for the extraction of feral stingless bee colonies from the forest, colony establishment and splitting, and the production as well as utilization of stingless bee products, among others are included in the package. It is characterized by its very low capital and labour-time investment requirement suitable for marginalized farmers.

RAC for Bicol at CBSUA offers facilities and installations for the communal use of local beekeepers, like rooms for seminars and discussions, colonies for hands-on training, an interpretation centre and a honey press extractor. Extension services and technical assistance to beekeepers in the region included assistance to beekeepers in marketing and promoting their products and meliponaries. Tangible outputs have been presented in scientific forums locally and internationally and published in reputable journals. Moreover, audio-visual materials are available in both print and audio on beekeeping activities.

Increasingly among graduate and undergraduate students in research possibilities to add value to the products obtained from the economic exploration of *T. biroi.* Students gain an appreciation for these bees thereby encouraging them to plunge into an enterprise like keeping stingless bees. To date, research is being done by students and faculty members of CBSUA to strengthen the technology on stingless bees. On a lighter note, its research outputs have won awards at the local and regional levels.

Brief History of Beekeeping in Bicol

In the past, only honey bees were only known to bee-hunters in Bicol who relied on honey as their source of income. Honey was harvested from *Apis dorsata* (putyukan) and *A. cerana* (liguan). The traditional practice of finding feral colonies in the forests, harvesting the honey crop and leaving the colonies destroyed was common. The Honey harvested was of poor quality since sealed and unsealed honeycombs were mixed during extraction. Extracted honey is placed inside pails with workers and brood found afloat when sold in the market. This situation paved the way to promote beekeeping and improve/develop the technology from production to processing.

In the late '60s, bee hunters had an abundant source of indigenous species of honey bees. Aside from the old practice of hunting bees from the wild, some of them hived the colonies and raised bees in their backyard applying the "trial and error" method of beekeeping. Zero-knowledge of the identity, composition, biology, behaviour of bees and quality of honey harvested was evident among the bee raisers. Moving on to the '70s and '80s, using twelve colonies of the native bees (*Apis cerana)* initial observations on the queen behaviour, bee space, hive size and other colony characteristics were done at CSSAC by one of the innovators (Prof. Ruiz). This was followed by the establishment of an apiary for instruction, research, extension and production activities of the college having both *A. mellifera* and *A. cerana.* However, due to a lack of support, the apiary ceased to exist. Two pieces of training for bee enthusiasts were conducted in Camarines Sur and at Uruan, Tiwi, and Albay. In Albay, the BIBERA (Bicolandia Bee Raisers Association, Inc.) was organized in 1983 and accredited in 1984 by the Minister of Agriculture Salvador Escudero III. Members of the said non-profit organization developed their farms which were eventually wiped out in 1987 due to typhoon Siang. At the Bicol University College of Agriculture and Forestry (BUCAF), 50 colonies of *A. cerana* were bought from Mr Josue Narce and maintained on the campus by Dr Mostoles for research and production purposes. Studies on bee behaviour, bee plants, pests and diseases, bee hunting and use of frames were done. Two years of bountiful honey harvest and generation of knowledge ended when the coconut wine (tuba) production on the campus caused a decline in colony strength resulting in wax moth infestation and the absconding of bee colonies.

In the early '90s, Prof. Ruiz from CSSAC revived beekeeping by putting up his apiary with colonies of *A. mellifera* which were used for hands-on training of CSSAC students. It was also during this period wherein Mr David General of the Ateneo de Naga College started an apiary for *A. mellifera* at Panicuason, Naga City. At BUCAF, beekeeping using *A. mellifera* was initiated after Dr Mostoles attended training at UPLB and bought colonies from the Royal Bee Corporation. Students of BUCAF and BUCAS conducted studies on this bee species. She researched the Biology and Control of the

Greater wax moth, the survey of existing apiaries in Albay. This paved the way to discover farmers at Magcasili, in Guinobatan, who have kept native honey bees headed by Mr Rodolfo Palconitin or *Tyo Opry*. In 1995, she transferred to CSSAC but still followed up on the undertakings in the said barangay. Collaboration with CSSAC was initiated to promote beekeeping despite the meagre financial support.

In 1997, beekeeping was revived with the joint efforts of Prof. Ruiz and Dr Mostoles who together with the CRS-AFP staff, and Entomology Major students attended the 3rd BeeNet Conference. She facilitated the attendance of Mr Palconitin and was formally introduced as an experienced native beekeeper. A poster was presented on the unique technology of raising the native bees, *Apis cerana* practised at Magcasili, Guinobatan as presented by Cervancia, Fajardo and Mostoles. The convention opened the doors for the beekeepers in the Bicol region. Popularity rose with constant visits by local and international beekeepers in the area. The RAM (Ruiz, Aquino and Mostoles) organized beekeeping training in the Bicol region focusing on *A. mellifera* in various parts of the region and even going as far as Mindanao. When Dr Mostoles was requested to train DAR beneficiaries on the culture of *A. cerana*, she recommends Mr Palconitin serve as the resource person and earned income out of his beekeeping activity. As such, the other locals in the barangay convinced of this simple livelihood ventured into keeping the native bees. The involvement of almost all the people in the sitio paved the way for the creation of the Mauraro Beekeepers Association (MBA) with the assistance of CSSAC. CSSAC adopted the barangay by serving as technical consultants and in return benefiting from the collaboration by serving as a laboratory for student and faculty research, extension activities and production initiatives in the area. Likewise, UPLB students and researchers carried out studies at Guinobatan and brought beekeepers from abroad to collect valuable data on our native bees. Individuals and groups trained by RAM frequented the said farms to buy colonies thereby making the rural folks earn additional income through beekeeping.

Particular attention was made to the possible benefits of the stingless bees (*Tetragonula biroi*) found on the house posts of Mr Palconitin, which claimed to be honey-forming. This bee species is abundant in their areas but they were not aware of its benefits and how they are managed. Together with Mr Palconitin, Dr Mostoles and Prof. Ruiz, as early as 1999, pursued the study of this endemic bee species including its biology and behaviour, cultural management practices and pollen sources. Hunting of colonies from the wild was done with Prof. Ruiz and Dr Mostoles providing financial support for the purchase of the feral colonies. These were maintained in the farmstead of Mr Palconitin and Mr Condino, which were the stocks used for the various exploratory research. Finally, in 2003, the culture and management of the stingless bees, a technology packaged and presented at the BCARRD Regional Symposium of Research Highlights won 3rd Place. They used the award from the said competition plus their funds to produce the VCD on the Cultural Practices and Management of *T.*

biroi which was presented at the 4th International Conference of the Asian Apicultural Association in February of 2004 and the BAPI convention in 2005. In effect, various linkages, nationally and internationally in research, extension and production were made. Research mushroomed on both imported and native bee species, with an emphasis on stingless bees. To cite one, the adaptability of the stingless bees to the Japanese-designed hive (OATH type) was done in collaboration with Dr Kazuhiro Amano of the National Institute of Grassland and Livestock Sciences, Apiculture Laboratory in Japan.

To date, interest in meliponiculture has mushroomed with the establishment of successful meliponaries in the region. The technology was introduced at Bulusan, Sorsogon by the Palconitins who initiated and were instrumental in putting up the Balay Buhay sa Uma by serving as technical staff and support. In the same area, community meliponaries were established through the SEARCA SFRT but sustainability was greatly affected by the presence of commercial meliponaries. At Casiguran, the Grajo's Bee Farm has blossomed through the efforts of a "father and son" team, Manuel and Leo Grajo. In Albay, aside from the TOP A Family Farm (owned and managed by Mrs Floreza Palconitin-Broqueza), Rafael's farm at Ligao City has expanded into the 3G Farms, a successful community beekeeping initiative headed by Henry Rafael, who had been a trainee of Regional Apiculture Center at CBSUA. Trainees from Masbate, Catanduanes and Camarines Norte have also started putting up their bee farms. In Camarines Sur, strong meliponaries are: Abad's Farm at Goa, MTB farm at Pamplona, Ecobees and U Farm at Pili, Billones' Farm at Nabua, Martirez Farm at Balatan, and many more small emerging meliponaries exists with CBSUA as instrumental in the promotion of the technology.

The Creation of the Regional Apiculture Center

With the establishment of the National Apiculture Research, Training and Development Institute (NARTDI) at DMMMSU, the beekeeping industry in Bicol was enhanced with CSSAC identified as the Regional Center for Apiculture. This was through the support of the late President Wilfredo G. Olano and endorsement by the Regional Development Council in 2006. Full implementation of the centre's programs started in January 2007. The CSSAC Bee Program is anchored to NARTDI's thrusts and goals mandated to work on the stingless bees. External linkages are being established with the various LGUs, provincial offices, GOs, NGOs and other line agencies. It implements the beekeeping RD & E priorities through the collaboration of these linkages to deliver the technology to various stakeholders in the region. It is inspired by its commitment to enhancing the sustainable productivity, profitability and competitiveness of the apiculture industry in the region, and will carry on with its tasks and obligations towards the realization of its goals and aspirations.

This paper will present the best practices on Stingless Beekeeping as a technology for marginal farmers not only in the Bicol Region but in the country. It will also present the stingless bee technology initiatives and directions of the Central Bicol State University of Agriculture.

Meliponicuture: Grassroots Technology Of Bicol

Meliponiculture:

The Stingless Bee Technology

Tetragonula biroi, the stingless bee locally known as "kalulot" is one good source of honey produced in the Bicol region. It is raised successfully in a truly indigenous method. The honey produced by this stingless bee had been claimed by indigenous people to be an effective cure for asthma and other respiratory diseases and is much more effective than the honey of *A. cerana*. This species of bees were observed to produce more propolis and pollen compared with other honeybee species (and *A. cerana* in particular, produces no propolis at all). Likewise, this species could be exploited for other useful products which help augment the income of the local bee raisers.

There are at least five species of *Tetragonula* (formerly called *Trigona*) that are known to occur in the Philippines. This genus was reported to be an efficient pollinator of crops such as coconut, mango, and cucurbits. Aside from their role as pollinators, stingless bees are pests of weak colonies of *Apis*. Unlike *Apis mellifera*, the imported honeybees and *A. cerana*, the native bees, much have yet to be studied on the culture and management of *T. biroi*. Most of the studies conducted were on the foraging and nesting behaviour of this species at Los Baños, mass rearing technique and analysis of the honey produced at Guinobatan, Albay.

At the Regional Apiculture Center (RAC), CSSAC, Pili, and Camarines Sur, the technology on the culture of these bees have been established as a product of research. Dissemination of the technology is being done through training, and attendance in scientific bee forums and its websites. Linkages, locally and internationally have been established.

Cultural Management of the Stingless Bees

New techniques for keeping stingless bees have been developed by the RAC with the collaboration of the local beekeepers from Guinobatan, Albay. A documentary on Cultural Management Practices is made available to budding meliponiculturists in almost all parts of the country, extending even to other countries.

Collections of feral colonies in Bicol showed different nest entrances and hive architecture as shown in Figures 84, 85. Nesting sites are dominantly ferns attached to coconut trees and trees (both forest and fruit trees). In Pilar, Sorsogon, a pili (*Canarium ovatum*) tree was observed to have 54 nests of stingless bees, indicating an abundance of feral colonies in the area.

The components of the technology are presented below:

Obtaining colonies. The colonies are acquired from parents, hunted from the wild, or purchased at 100-200 Php per colony from established beekeepers. At present, a strong colony is now being sold at Php 800-2000, depending also on the hive material. Hunted colonies are found on trees embedded in ferns. Removal of the colony is done by carefully cutting the hive, lowering it down slowly maintaining the position to prevent the honey from dripping into the brood combs, later leading to brood death.

Figure 293 Nest architectures of feral colonies collected in Bicol.

Figure 294 Hive entrances of feral colonies collected in Bicol

Hiving a colony. Some use halved coconut shells placed on top of the side of the entrance holes. Shells are fixed by bees using propolis. Others leave the colonies in the wild undisturbed. Hived colonies are hung on a tree using a tie wire while others are set on 1-1.5 meter posts. The Japanese-design hive (OATH type) is recommended for propolis production and pollination jobs. The innovation of the OATH type as well as Henry Rafael's apartment type was found to be accepted by bees. The KITH type (round type) is being tested. Shown in Figure 81 are the different modifications of the stingless bee hives.

Locating the colonies. Hives must be placed in a meliponary with abundant sources of pollen and nectar, chemical-free and predator-free. A tree can serve as shelter from overexposure to sunlight and rain. Areas with fruit trees, coconut, flowering weeds and crops are suitable for the bees. If placed on posts, distance from the ground must be considered for ease of inspection and harvesting. Mixing with *Apis* poses a risk of creating problems with disease drifting, robbing or competition.

Colony shelter. Traditionally, anahaw leaves, coconut fronds, GI sheets, and plastic materials are used to prevent water seepage through colonies. Innovations such as the provision of GI sheet hats or triangular housing allow protection from adverse environmental conditions (*Figure 82*).

Feeding of bees. Artificial feeding is not practised but the stingless bees are fed with fruit peels and spoiled honey of the native bees. There is an abundant source of pollen and nectar from coconut, fruit trees, forest trees, weeds such as *Mimosa* and *Wedelia*, and even *Jatropha*. Studies on pollen sources, bloom patterns, pollination and adaptability to different ecosystems of *Tetragonula* had been studied.

Harvesting of honey and pollen. Pollen and honey pots are harvested from coconut shells by pricking the "eye" of the nut. A transparent but thick substance on the stick indicates a honey pot-filled shell, yellowish granules for pollen and white liquid for the brood. Pollen and the honey-filled shell are detached from the hive and set aside for extraction. In the OATH, KITH and Apartment types, the opening of the entire hive is done which could pose damage to the brood combs. After harvesting, colony division is done.

Figure 295 Different types of hives developed and evaluated

Colony division. Determination of the colony strength must have at least 38 brood combs (17 larvae, 21 pupae), covered with pollen and honey chambers and at least 2-4 queen cells present. The division is achieved through the swarming or eduction method. Swarming is the natural division wherein the swarm is collected and hived. Eduction is an innovation achieved by equally dividing the colony once two entrance holes are evident. An artificial entrance hive could be provided using a 0-5-3 cm pipe.

Extracting honey, pollen, and propolis. Extraction of honey from pots is done by scooping with a spoon and placing it in a double nylon tulle, honey is squeezed or pressed. A honey press extractor designed at the NARDTI is an innovation used in extracting big amounts of honey harvests. Extracted honey was poured into a basin and then double-filtered into a cheesecloth. Extracted honey is let to stand for 24 hours for impurities to float. After which, honey is placed in glass (wide-mouthed) jars for the bubbles to be released. The bottling of honey is done after two days. Pollen is scooped individually from each pot, transferred into a clean container and dehumidified. An improvised dehumidifier is used to reduce the moisture of pollen before packaging. Propolis is obtained from the shell lining the pollen and honey pots, from the batumen of the colony. It is washed using tap water, air-dried and refrigerated. Ethanolic extraction is done for two weeks. Harvesting of extracts is done, filtered twice and bottled.

Figure 296 Some of the colony shelters the stingless bees in meliponaries.

Honey processing and product utilization. Technology generated manifested by the products produced for cosmetics and medicinal supplements derived from honey, pollen and propolis from stingless bees. For cosmetics, products are Bee Propolis Cream, Bee Propolis-Honey Cream, Bee Propolis-Honey-Pollen (PHP) Cream, Shower and Bath Gel, Propolis Soap and Propolis Shampoo, Propolis astringent, and Propolis Balm. New products from a project funded by DA BAR are the Improved Bee Propolis Liniment, Tincture of Propolis and Bee Propolis Body Scrub. For food, the Honey Mead/Wine, Lukot Pure Honey, Lukot Honey Macaroons, Lukot honey-coated Pili and Peanuts, Propolis Soap, Lip Balm with propolis, Shampoo bar and many more.

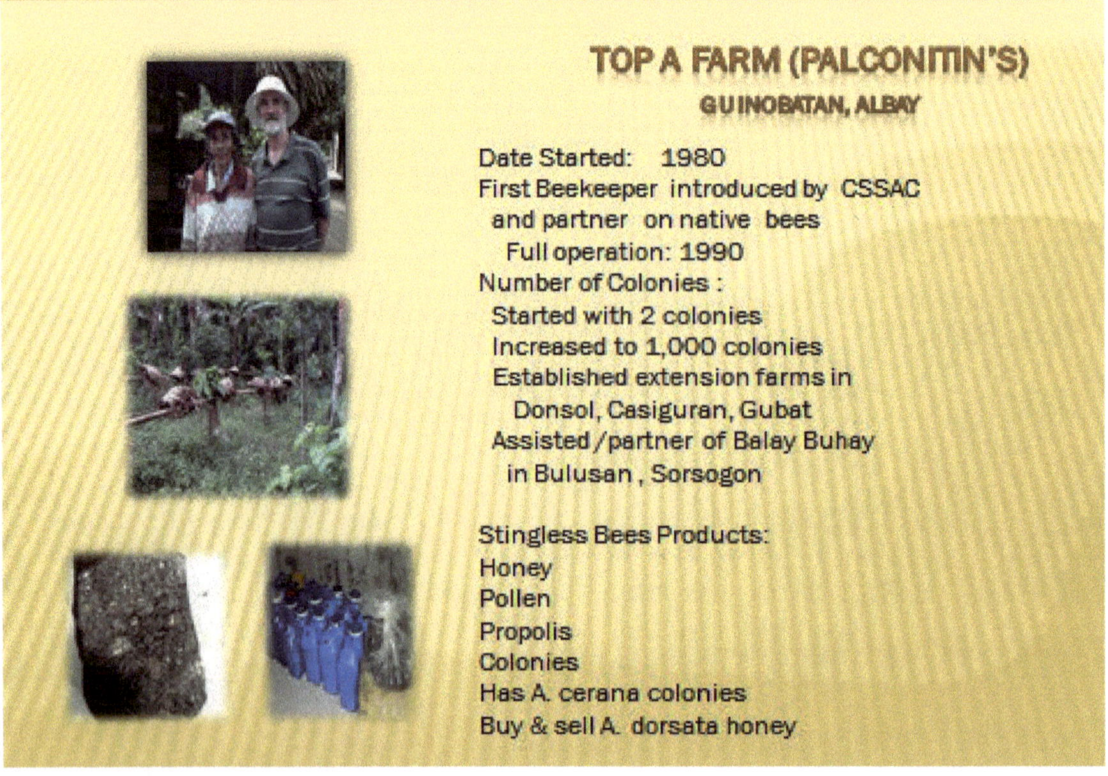

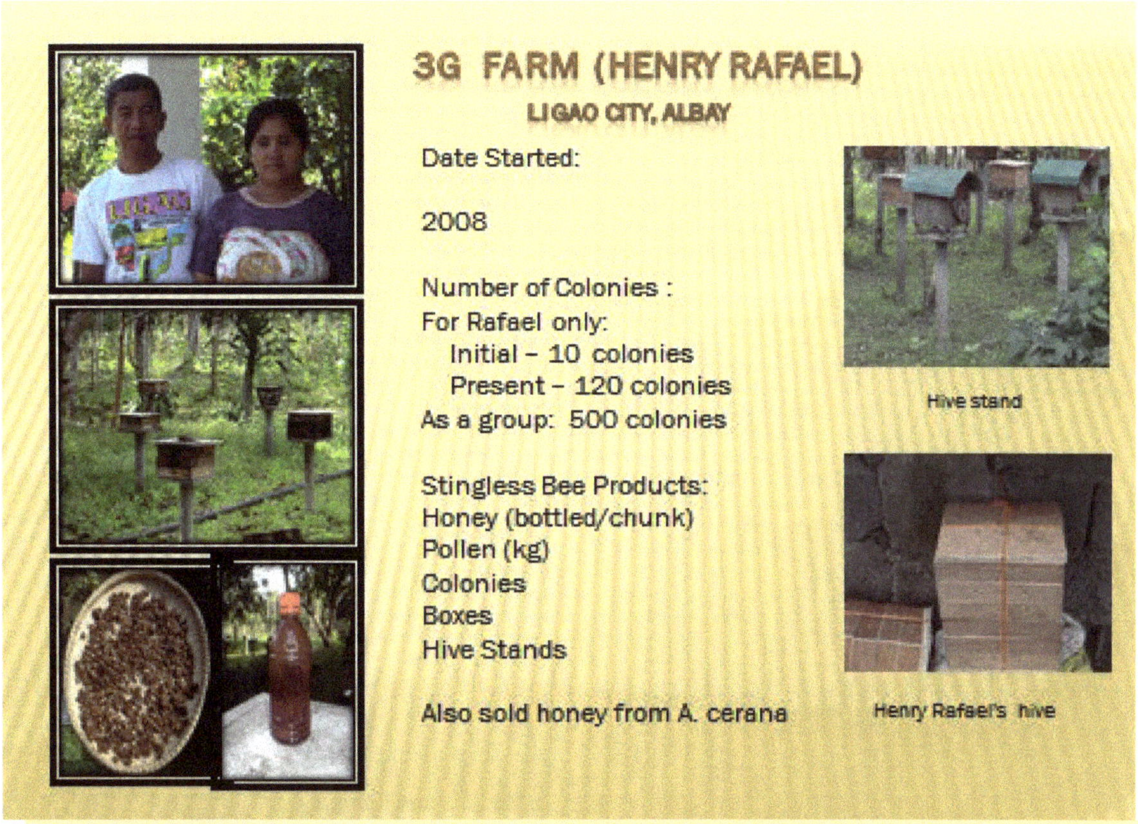

Figure 299 Meliponaries in Bicol with a market survey. – 3G Farm (Henry Rafael) Ligao City, Albay & Phi Apiary Castilla Sorsogon

Figure 300 Meliponaries in Bicol with a market survey conducted. – Ecobees & U Farm Enterprise, Pili, Camarines Sur.

Meliponary Establishment. Apiaries using stingless bees, called meliponaries have been established and monitored. Meliponaries exist in different areas assisted by the CBSUA: In Camarines Sur, Parish of St. Dominic (Gainza), Abad's farm (Goa), Belmonte's farm (Naga City), Curry, RAM Apiary (Pili), Gawad Kalinga (Lagonoy), Joseph's Farm (Naga City), Hidalgo's farm (Calabanga), Sipocot campus, Carreras' farm (Tigaon), Wood's farm (Panicuason), Iriga City, and Del Gallego. In Albay, Palconitin's apiary (Guinobatan), Condino's farm (Guinobatan), and Rafael's farm (Ligao City). In Sorsogon, Sorsogon State Colleges (established as the Provincial Apiculture Center). In Masbate and Camarines Norte- in collaboration with the Department of Trade and Industry, Batuan and Based, respectively, have started apiaries for CARP beneficiaries. Some of the apiaries were damaged during typhoon Reming but will be recovering shortly. Assistance was also provided to the Catanduanes State University which was also established as a Provincial Apiculture Center by CBSUA. To date, commercial meliponaries are found in the Bicol region. Dissemination of the technology has extended to various parts of the country through CBSUA, to wit: established farms in Pampanga, Iloilo, Tacloban, La Union, Ilocos Sur (Candon Cty), Pangasinan and Sanchez Mira in *Trigona* from 2001-2006 with an income derived from the sales of honey to as much as 530,000. Some of the meliponaries which are commercially producing products are shown in Figure 297 through Figure *302*.

Figure 302 Meliponaries in Bicol with a market survey conducted. – FR. Barcelona's Apiary, Pilar,

Sources of Income

Farming		Yield/Values			
Crops Planted	Ave.no/hill	1990	1993	1995-2000	2001-2006
a.coconut	1.0 ha	27,000			
b.banana	30 hills	4,500			
c.anahaw	50 hills	6,200			
Total		37,700			
Other sources					
Beekeeping					
1. Apis cerana					
Colonies	47 colonies	0	79 (P 12,000)	120-140 (968,000)	147(700,000)
Honey(4 mos)	4 gal/mo	4,800	72(7,000)	80(12,000)	100(15,000)
2.T. biroi					
colonies	2	-	10	60	2160(500,000)
Honey	0	-	-	-	150(30,000)
Total		42,500	17,000	980,000	1,245,000

Table 1 Tabulation of Annual Source of Income from Farming Yields

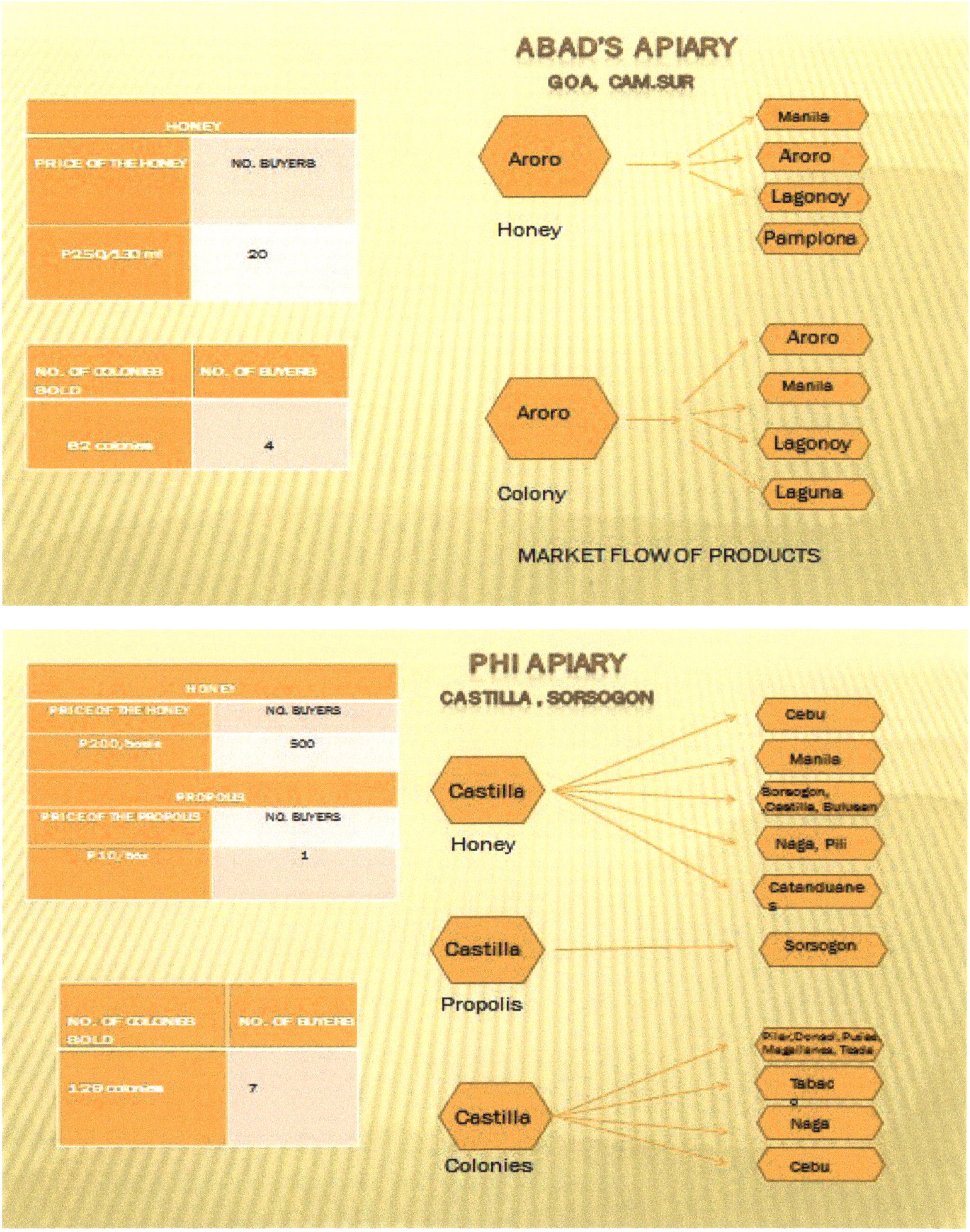

Chart 8 Market flow of stingless bee products in selected meliponaries. (Abad's Apiary & Phi Apiary)

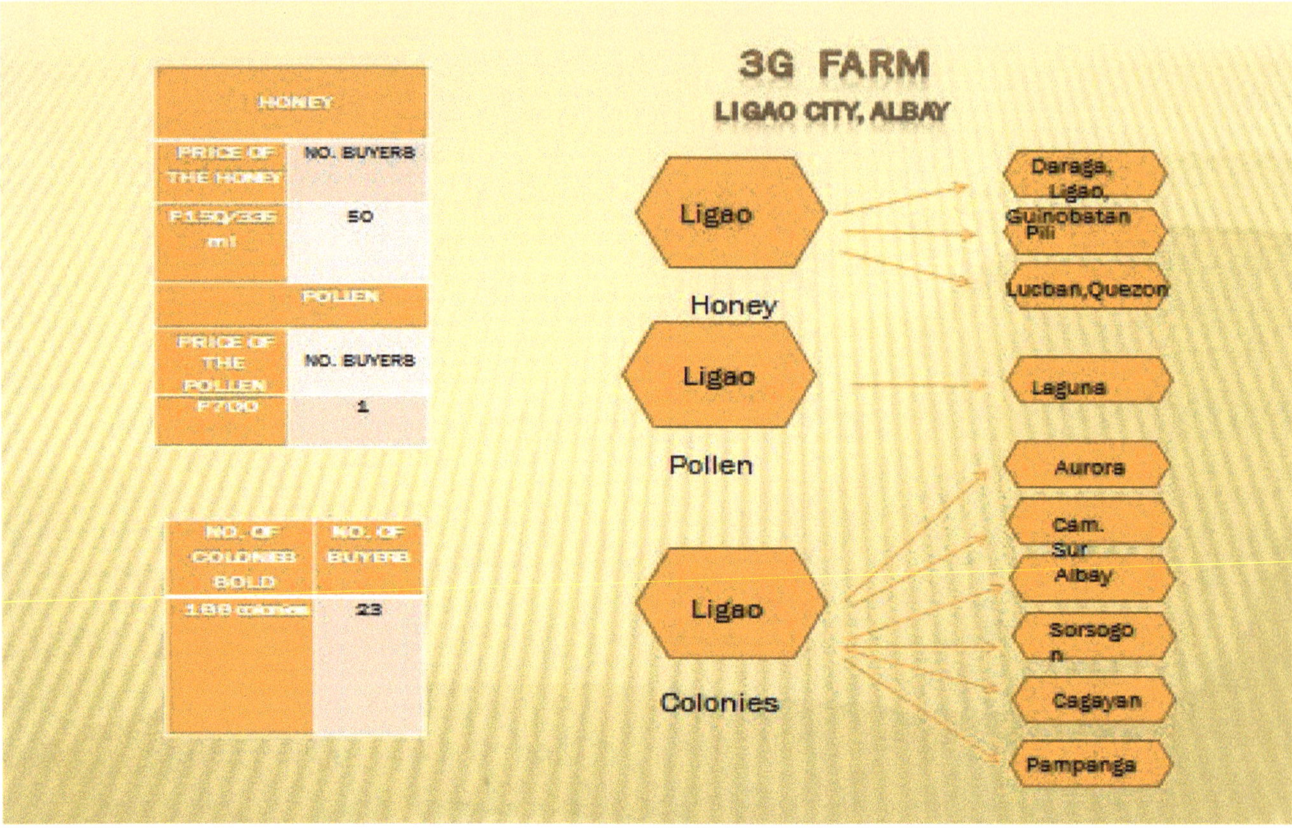

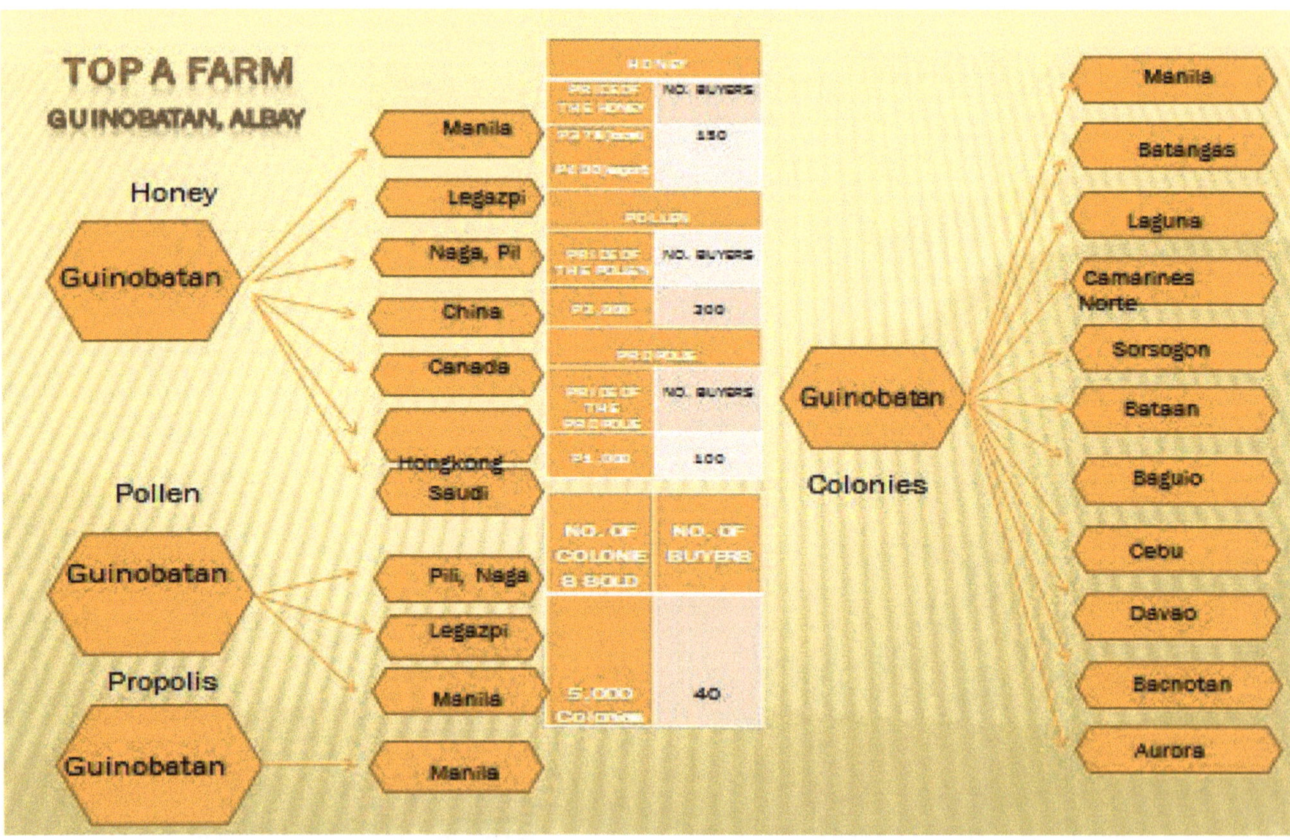

Chart 9 Market flow of stingless bee products in selected meliponaries. (3G Farm & Top A Farm)

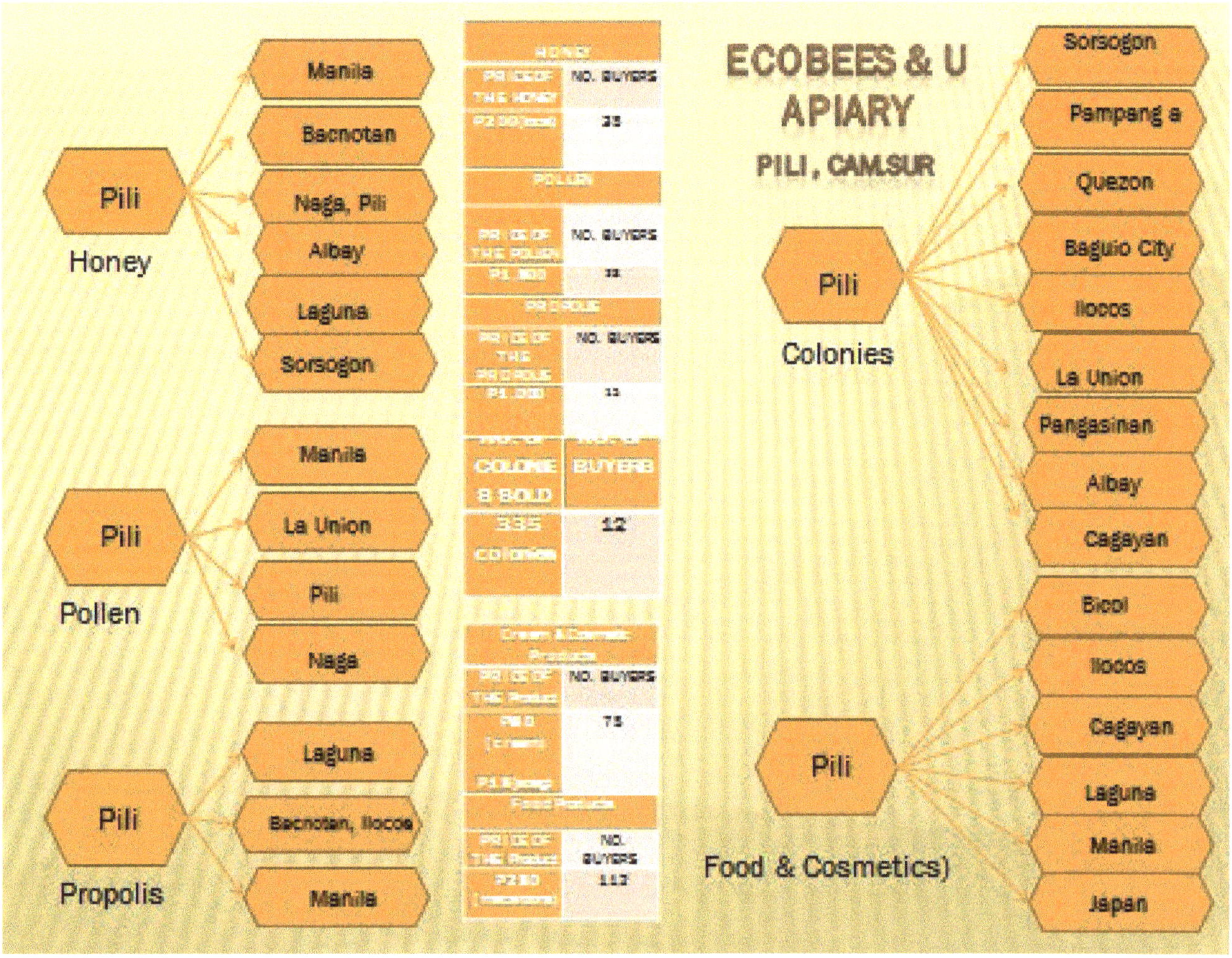

Chart 10 Market flow of stingless bee products in selected meliponaries. (Ecobees & U Apiary)

Uses /Areas of Application:

Meliponiculture is suitable in a wide variety of areas, however, it would be more sustained in areas where there is a sufficient source of pollen and nectar for the bees. It is applicable even in marginal areas as well as with indigenous peoples. Since feral colonies are found in the forests, this technology could likewise be a means of conserving the biodiversity of our forests. It will also serve as the indicator of pollution in the place where it is cultured since bees will not thrive in polluted areas. The initial application of the technology was at Magcasili, Guinobatan wherein the main source of income by the community was farming and labour in a winery and other businesses in the town proper. At first, only two or more families were adopting the technology, but after a year or so, with the popularity of beekeeping and income coming in, the entire community finally ventured into beekeeping. As such, it became a community enterprise which was strengthened with the creation of the Mauraro Beekeepers Association which paved the way for organizing the Bicol Beekeepers Association. At present, the technology has radiated to other provinces in the region. Colonies of stingless bees are now being imported and cultured in Laguna, Cavite, Quezon, Aurora, Tuguegarao, La Union and Benguet provinces.

Advantages of Meliponiculture

Beekeeping using stingless bees has advantages over other apiculture endeavours.

1. It is simple and flexible. The technology can easily be adapted by local people since it first emanated from the locals. It has been packaged in such a way that any new or budding beekeeper could easily follow. Since the bees are highly endemic in the region, adaptability to a certain location or area will not be a hindrance to sustained production. There are no dreaded diseases or pests associated with the bees, thus, control is unnecessary.
2. It is eco-friendly. They do not cause harm to humans since they are stingless. Their products and by-products such as honey, pollen (in the form of beebread), propolis, and wax are beneficial to man and animals. Stingless bees are indicators of pollution and indirectly promote biodiversity conservation. With the knowledge that flowering plants, fruit and forest trees as good sources of pollen and nectar for the bees, cutting trees (discouraged) and planting bee plants would be encouraged leading to environmental conservation. Enhanced pollination of crops could be achieved due to their small size and polylectic behaviour, and symbiotically, the bees benefit from this by an increase in by-products.
3. It requires low capital to start up a meliponary. An ordinary farmer can afford to culture this bee species. Most of the materials needed to keep these bees are locally available and at times considered wastes in production.
4. It is backstopped by research which could lead to further improvement of the technology. CBSUA is researching systematics and distribution, bee biology and behaviour, pollination, by-product processing and utilization, marketing, conservation and ecosystem management, and various extension activities.

Moving Forward

In the pursuance of its mandated goals and objectives, a full implementation for the past two years (2007 and 2008) has generated as well as packaged the Technology on Beekeeping using *Tetragonula biroi* which is now in full commercialization. Funds to maintain and operate the centre have been derived from income generated in training and initiative from its members.

342

The centre and its programs are consistent with the mandate on academic program offerings, research, development, extension and production activities of the institution. The significant accomplishments of the RAC in the various areas are presented below:

A. Academic Program

1. Beekeeping was taught to Entomology students in the course Entomology 170 (Industrial Entomology), a subject which is a regular offering for students taking up Bachelor of Science in Agriculture, majoring in Entomology. It is also an elective subject for students interested to learn the basics of beekeeping.
2. Beekeeping has been utilized as an entrepreneurial activity among the BS Agribusiness students through their advisers who trained under the centre.
3. Beekeeping served as the practicum by students in Entomology and Plant Pathology which was offered from 2008 up to this year.
4. Beekeeping served as an extension, research and instruction activities of the CBSUA Entomological Society.
5. Support students in the conduct of their thesis both at the undergraduate and graduate levels.
6. Documents of the RAC contributed to the BSA Accreditation, Center for Development and NUCA. Likewise, it has supported the graduate programs, Environmental Science and Agroforestry Programs' Accreditation.
7. Scholarship provided to a faculty member to specialize in the taxonomy of stingless bees.

B. Research Program

The research activities focused on bee identification, biology and diversity of stingless bees; cultural management practices, pest and diseases, processing and utilization of bee by-products and socio-economic impacts of beekeeping. Various studies in line with these were conducted as well as on other species. Completed researches include Processing and Utilization of Bee Products and By-products: Development of propolis cream and astringents for dermatological purposes, Processing and Utilization of Bee Products and By-products:
Study 1. Antimicrobial and antibiotic properties of propolis from *T. biroi*; Study 2. Control of dermatological diseases using propolis from *T. biroi*, Morphometric analysis of stingless bees in Bicol, Status of selected beekeepers in Camarines Sur and Albay, Bee R&D Program (packaged with 3 studies), Pollen sources of stingless bees about floral bloom pattern in selected ecosystems in Camarines Sur, Pollination of *Jatropha* using native bees, Adaptability of stingless bees to different

types of hives, Performance of stingless bees in different hives under different ecosystems, Abundance and distribution of stingless bees in Bicol, Pollen Sources of Stingless Bees in Bicol, Pollination of cucumber, sunflower and pigeon pea using stingless bees, Floral Phenology of Bee Pasture Plants in Selected Meliponaries in Camarines Sur, Community–Based Approach to Sustainable Utilization, Management and Development of Beekeeping in Sorsogon, Bioinformatics of Stingless Bees (Meliponinae: Hymenoptera) in the Philippines.

Other research was conducted by students on the nests and nest architecture of stingless bee species, marketing of bee and bee products, the morphology of bees, production of beebread and propolis of stingless bees as well as pollination of different crops using stingless bees. Different ecosystems were also explored in the presence of stingless bees.

Research outputs were presented in various events which won awards at the local, regional, national and international levels. The packaged technology on Meliponiculture was featured in the 50th Anniversary of the Philippine Association of Entomologists during the PMCP convention in 2013.

C. Extension Program /Community Engagement

Promotion of the stingless bee technology was done through training, forums, exhibits, and the establishment of demonstration farms. No less than 50 training were conducted not only in the Bicol region but in other parts of the country. Assistance was given on the following: the establishment of apiaries for new beekeepers including the purchase of colonies; apiary establishment by the Sorsogon State College, Catanduanes State University and Cagayan State University, and Marketing of products of beekeepers.

Demonstration farms have been established in private organic farms through a memorandum of agreement. The Crop Diversification Project, tagged as Bahay Kubo ni Manay showcases Meliponiculture and other technologies generated by the university interfacing with instruction, research and production.

D. Enterprise/Production Program

Products of stingless bees and nuclei (starter colonies) are being produced as a component of the Entrepreneurship Initiatives of the university. Raw honey is sold either wholesale or retail. "Nucs" or nuclei which are starter colonies for would-bee meliponaries are sold at a varying price determined by the strength of the colony and the type of hive material used.

Continued production of IEC has been done with the following: RAC Information Bulletin, Honey Apitherapy, and Benefits of Beekeeping. Bee Pollen, Bee Propolis, Abundance and Distribution of Stingless Bees in Bicol, Food and Cosmetic Products using Stingless Bee Products, Cultural Management Practices of Stingless Bees in Bicol, Pollen Sources of Stingless Bees. DVD/CD on the Cultural Management Practices of the stingless bees in Bicol and Beekeeping in Bicol has been produced and distributed to trainees.

References

Batista, M., Ramalho, M., & Soares, A. (2003). Nesting sites and abundance of Meliponini (Hymenoptera: Apidae) in heterogeneous habitats of the Atlantic Rain Forests, Bahia, Brazil. *Lundiana 4(1)*, 19-23.

Bradbear, N. (February 2004). Bees and the Environment. Proceedings of the 7th Asian Apiculture. *Association Conference.* (pp. 23-27). AAA- BEE Net-FFTC.

Mostoles, M., & Ruiz., R. (2010.). "Meliponiculture: Grassroots Technology Bicol Regional Apiculture Center.". *ICHER 1(1)*, 183.

Mostoles, M., Bien, R., & Lanzuela, F. (2010). Stingless bees pollen and propolis and their production in Camarines Sur". *Philipp. Ent. 25(2)*, 202.

Mostoles, M., Buenaagua, R., Pasiona, L.C., Rosario, D., A.B., B. R., . . . Espejo, M. (2015). Performance of stingless bees (*Tetragonula biroi* Freise) in Different Hives under Different Ecosystems. *The Philipp. Ent. 29(20)*, 123-135.

Mostoles, M., Buenaagua, R., Rosario, A. D., & Jovillano, H. (2014). "Food and cosmetic products using stingless bees ". *Philipp. Ent. 28(2)*, 208.

Mostoles, M., Monforte, J. V., J., Reorizo, R., Briquillo, J., & Broqueza, F. P. (2014). "Production, commercialization and utilization of stingless bees in Bicol". *Philipp. Ent. 28(2):*, 216.

Mostoles, M., Pasiona, L., Buenaagua, R., & Reabad, D. (2010). "Adaptability of stingless bees in different hives". *CBSUA Journal of Research. 8*, 7 -11.

Mostoles, M., Pasiona, L., Buenaagua, R., & Reabad., D. (2010). "Adaptability of stingless bees in different hives". *Philipp. Ent. 25(2)*, 201.

Mostoles, M., Pasiona, L., Buenaagua, R., & Rosario, A. D. (2014). "Pollen sources of stingless bees and bloom pattern in selected meliponaries in Bicol". *Philipp. Ent.28(2)*, 225.

Nicolas, A. (2008). "Geographical distribution, morphological and colony characterization of stingless bees(Hymenoptera:Apidae: Meliponini) in selected provinces of the Bicol Region". *MS Thesis, CBSUA, Pili, Camarines Sur.*

Salazar, G. (2007). "Pollination efficiency of stingless bees (Trigona biroi Freise) on cucumber (Cucumis sativus) and bell pepper (Capsicum sp.)". Dissertation, Camarines Sur State. *Dissertation, Camarines Sur State Agricultural College, Pili, Camarines Sur.*

Samejima, e. a. (2004). The effect of human disturbance on a stingless bee's community in a tropical rainforest. *Biological Conservation 120*, 577-587.

Slaa, E. J., Sanchez, C. I., Malagodi-Braga, K. S., & Hofstede, F. E. (2006). Stingless Bee in Applied pollination: Practice and perspectives. *Apidologie 37*, 1-23: 293-315.

CHAPTER 19 – ALBAY PROVINCE

19.0 **Etymology**: Albay - From Spanish Albay, as in al ("the") + baye ("bay"), or in longer variant names, La Baye la baye ("The Bay of Baye" or Bay la Bay ("Bay of Bay") as in Olivier van Noort's 1602 Map of La Baye la baye (Albay), where the first baye or bay is an old variant spelling of Spanish bahía ("bay (geographical feature)"), but the second baye or bay refers to its previous derivation from Spanish Al Baybay, which itself came from Spanish al ("the") + Bikol Central baybay ("shore"), where finally, it was simplified or shortened to just its present form to do away with any confusion.

19.1 Cristina Imperial, Ahrdail Subdivision, Tagas, Daraga, Albay.

In Jube 2020, a typhoon was headed to Albay if tracking doesn't change... Ms, Cristina said, "this time I am thinking of just extra securing hives and no longer removing them for safekeeping in the garage...any ideas on how to secure them better?!"

AHJ assessed that the stand is too high to be stable and asked how many colonies she has like that.

She said that she has four colonies like it, so AHJ suggested that she should consolidate all of them on a lower (waist-high) more stable stand anchored to the ground. The boxes are affixed with brackets to the stand.

Image 66 The boxes are affixed with brackets to the stand.

Ms. Cristina replied, "Ok, I will see how we will go about it and also monitor the kph of the typhoon."

Image 67 I discovered this while having my house repaired

She continued, “I discovered this while having my house repaired, I hope I did the right thing, should I slice open the PVC pipe? I just transferred the PVC inside the box at the moment.”

AHJ then answered, “Looks appropriate to let it develop in the box first. Though the box need not be so tall... half the height would have been good enough. No need to put back the honey-soaked propolis cos it will attract ants and even roaches. Was it originally part of your roof eave drainage or was it part of some lighting fixture? “

Ms. Cristina replies, “Yes, it’s part of the roof drainage... Ok, will remove the top box, remove the propolis and honey. The whole roof flew away during typhoon Rolly last November! It just looks ok from that angle.″

Image 68 We found some of our roofing on the beach after the typhoon!

Maam Cristina owns a Pili nut Farm and keeps stingless bees to utilize them for pollination with the prospects of increasing their yields. Honey produce would be an added yield/ She attended the advanced course in Lipa City in March 2020. One of the discussions with her was about the presence of pollen in harvested honey. It was expressed that pollen content would enhance the potency of honey. I pointed out that the unequal amounts of pollen in each batch of harvest do not keep the potency consistent. It would be better to separate the pure honey from pollen content first and then add the appropriate percentage consistently over all the subsequent batches. This way will allow a better and more precise evaluation of the properties of the honey’s medicinal value.

19.2 Henry Rafael in Ligao City, Albay

Interviewed by Abu Hassan Jalil

Good day Sir, I was told by Apo Mariano (Ace Lonzaga) that you are gathering hands-on experiences on our stingless bee (Philippines). I will be glad if I can also share some of my experiences with it.

Q: Yes thank you for responding. Do you have a stingless bee farm? How many hives do you have? What species do you keep? Do you have pictures of your farm and what method of honey harvesting do you practice?

A: Yes I have a stingless bee (farm). I have today 150 colonies of *Tetragonula biroi* (according to Dr Dulce Mostoles. Dr Clare Baltazar was the systematist who identified *T. biroi*. Pers. Comm Cervancia 2016)) (see *Figure 89*Figure 89)

A: All my colonies are now inbox hives (I called them condo hives) (see *Figure 89*). Our method of harvesting honey is done manually (opening the box and then scraping the honey) (see Figures 90Figure 90 b, c & d)

Q: How many years have you kept the bees... where do you obtain the initial hives? What is the percentage of failures you encounter? How did you obtain the original hives?

A: I started keeping T. biroi almost 10 years ago. A friend gave a colony to me. From there, I started to study/ explore the beauty of these bees. I hired a friend to hunt for me some colonies (10) from the wild.

Q: In an area of 1 acre in the wild, how many hives of T. biroi exist?

Do you use the coconut shell method?

A: Percentage of survival...transferring the colony from original hive to box hives, up to today almost 90 % success.

I do not have an idea of how many colonies of T. biroi per hectare.

But since our place is deforested already, it is seldom seen in the wild.

Q: Where is your location exactly?

A: My farm is located in Paulog, Ligao City, Albay Bicol Philippines

That is my farm location- a 3G bee farm 470km South-East of Manila (see *Figure 90* a)

Image 69 150 colonies of Tetragonula biroi c) d) e) condo hives

Image 70 a) farm location b) manual honey harvest c) harvested honey d) other products e) stock of hive stands f) dragon fruit plants g) pitaya flower visited by stingless bees h) i) brood cells j) k) l) bees outside hive

Q: What is the production rate of Honey you are getting on average from your 150 colonies? Where do you market them and at what price, please? What are the challenges you face in your *T. biroi* meliponiculture?

A: My 150 are not all strong colonies yet. Maybe 50% of them are the strong ones. Because some of the strong are divided/split, sold to parties interested also to keep bees. Right now we cannot serve the growing demand for colonies (cultured in boxes). I can harvest an average of 5 kg of honey monthly. Market? Locals and nearby residents are my buyers, mostly for personal consumption.

Challenges: Growing demand for colonies and honey locally and in other provinces. Also lesser support from our government..e.g. promotion of technology. I am selling it for Php 300 per kg. a complete of my colony is priced at Php 800 farm gate price

Our honey is sold at Php 300 /kilogram a complete set of a colony in a box hive cost Php 800

Also, a common problem here is weather especially from July to December, with typhoons averaging 20 annually.

Q: Do you find an increase in yields in your dragon fruit with the biroi pollination?A: I have just 2 posts of dragon fruit, red and white only. 2nd year of bearing fruit only. I observe that almost all visible pollens of dragon fruit are collected by bees (*biroi* and *cerana*)(see *Figure 90*Figure 90 f & g)

Note* Stingless bees do not pollinate dragon fruit. They just gather nectar and pollen. (Pers Comm. Cervanvia 2015)

Q: What do you call meliponines in your dialect?

A: Philippine local name for stingless bees: *Kiyot / Lukot / Lukutan / Kiwot / Libog / Kelulot*

19.3 New Bee Farm Development in Albay

Visited Guinobatan to view the newly cleared area for Bee Farm Development. Albay District at the foothills of Mt. Mayon

The group that wanted to develop the piece of land brought us there to seek some advice from us on the "Beescape" planning. However, when I asked for the Masterplan to provide some planning input, they did not have one. So, we simply browsed around the cleared area and discussed simple matters along the way around. It was kind of muddy and I was not too keen (Figure 92).

Figure 303 Bee Farm Development. Albay District at foothills of Mt. Mayon

Figure 304 Although it was cloudy, we did catch a glimpse of Mt. Mayon

Although it was cloudy, we did catch a glimpse of Mt. Mayon on the way back to Central Bicol. We stopped along the way for dinner while discussing further consultancy for the Beescape development of the proposed bee farm project in Albay.

CHAPTER 20 - SORSOGON PROVINCE[27]

20.0.1 Etymology: Sorsogon - Hispanicized form of ***Sorsogon***, Bikol adjective meaning "wadable", derived from the root word *sogsog*, which means "to wade" or "to ford"; a wadable river, for example, is known as *salog na sogsogon*. The toponym originally referred to the settlement on the banks of the Salog River, which became the capital of the province upon its establishment in 1767.

20.0.2 Getting there: The more decent airport in Sorsogon is Legazpi and we took a Cebu Pacific flight from Manila to Legazpi. It was an early morning flight, and we had to leave the SEARCA guesthouse at 1 a.m. to catch the 4 a.m. Legazpi Flight from NAIA.

20.1 Philippines Honeybees Industries (PHI) Pilar

Philippines Honeybees Industries (PHI) Bibera Research Training Centre in Pilar, Sorsogon, Bicol Region, Managed by Mr. Domingo. We stopped by to observe progress on Kiyot hive propagation by the Coconut shell method from Fern hives obtained from the wild. Coconut shell hives are propagated under the cool protection of fan palm attap roofing, neatly arranged by traditional methods.

Figure 305 Philippines Honeybees Industries (PHI) Bibera Research Training Centre in Pilar, Sorsogon, Bicol Region, Managed by Mr. Domingo.

[27] By air, the nearest airport is Legazpi City in Albay Province and we took the opportunity to visit some historical sites there before heading to our destination in Bulusan, Sorsogon Province.

20.2 Leo Grajo of Sorsogon Philippines (interviewed by Abu Hassan Jalili)

After Albay we visited Grajo's Beefarm in Sorsogon.

Figure 306 a) b) & c) Tetragonula biroi nests d) Brood chamber with involucrum e) f) g) h) i) j) & k) coconut shell hives l) farm location

I am compiling some info on meliponiculture in your area. I have a list of questions and If it is ok with you I would appreciate it if you may describe patiently each question one by one.

Q: Do you have a stingless bee farm and how many hives do you have?

A: Yes, an estimated 450-500.

Q: What species do you keep?

A: *Tetragonula biroi* are the most common. But based on our observation we have different types of nest entrances. With your help, we may be able to name the other species/ variety that we have on the farm. (see *Figure 91*Figure 91)

Q: Are you practising the coconut shell hive method?

A: We are using coconut halves. It's one of the successful methods in the splitting/division of our hives. If we have a new wild colony we let it sit for a couple of months on the farm before opening and putting the coconut shells/the box on top of the comb. The bees will eventually fill the empty coconut shell/ box with a comb but occasionally this shell/box becomes food storage for pollen honey. This shell with a comb will be extracted/ divided into a new colony. (see *Figure 91*)

Q: What method of honey harvesting do you practice?

A: Manual extraction. We poke the coconut shell/box with a fork and invert the shell/box so the honey will drip into a container and strain the honey to separate the pollen batumen, etc.

Q: How many years have you kept the bees...

A: We started in October 2010. I remember I purchased more than 250 plus colonies from the wild. It was a procurement rush of mother colonies without deep knowledge of the management of stingless bees. A combination of excitement, a trial which ended in high mortality, leaving me quarter colonies in the next year. I worked outside the country so I left my colonies unattended and once a year, I come came to the Philippines to have a vacation/ splitting of bees but eventually, the rate was stagnant. It was only in 2013 that my sister manages the bees and slow increase the colony production.

Q: Where do you obtain the initial hives?

A: We just hunted within the local area.

Q: What is the percentage of failures you encounter?

A: On the first batch of procurement it was 75% mortality. These are the factors that contributed to the mortality:

Lack of knowledge in the extraction of wild colonies,

Lack of proper management on the splitting of hive/division.

Figure 307 Hives amidst vegetation for shade. Newly boxed hives

Figure 308 Hive racks and a Hexagon Bee House.

Figure 309 Initial Transit and transfer area for Fern Hive splits.

Figure 310 Fern with topping box trial.

Infestation occurred. (see *Figure 94Figure 94*)

But starting in 2013 lessens the mortality by learning the proper management method thru publication from Latin America. Now we can say that still have less than 10 per cent mortality.

Q: How did you obtain the original hives?

A: we asked local people/ honey bees, and honey hunters, to get mother colonies in the wild and buy from them.

Q: Where is your location exactly?

Typical pest infestation.

Figure 311 A. Side view of Hermetia ilucens (BSF) with larva; B.Underside of Black Soldier Fly; C. Parasitic BSF Larva infestation in nests; D. Haptoncus luteolus; E. Pollen Beetle (HL) infestation in Box hive; F. Side view of Reduviidae; G . Assassin Bugs Reduviidae; H. Cockroaches.

A: We are located in Region (the Bicol Region Our town is Called Casiguran (Coordinates: 12°52'N 124°01'E), province of Sorsogon. The farm is located specifically in Barangay Tigbao, Casiguran Sorsogon. Philippines (see *Figure 91*Figure 91 *l*)

Q: What is the production rate of Honey you are getting on average? Where do you market them and at what price?

A: We don't have much data on the exact honey. Our main goal is to mass propagate our stingless bees. But we split/ divide the colony into two seasons. From May to June that is the first season split and followed by October to November. Our main target is to produce at an exponential rate per year, but as experienced, it's a fluctuating rate. The major factor that contributes to unstable production is our climate. Our province is frequently visited by a typhoon which contributes to mortality, insufficient food supply and slow growth of colonies. Last December our town was devastated by a typhoon and the honey that is only produced last June was only 3 gallons and 8 kilos of dehydrated pollen.

We only select the strong colony to be split or divided that's why we only have 40 new colonies last June.

The harvested honey is marketed locally.

Figure 312 a) Bottled Bee Pollen b) Stingless bee Honey in jars c) freshly harvested bee bread

Stingless bee Honey Php 200 (200 Phil Pesos per 400grams) *Figure 95 b*

Bee pollen: Php 200 (200 Phil Pesos per 250 grams) *Figure 95 a*

Occasionally my sister produces tincture and salve which are sold also locally.

Q: What are the challenges you face in your species in Sorsogon?

A: The typhoon is the most destructive natural factor in stingless bee production but thru the years we try to cope with and design a stingless bee hive box that is typhoon-proof. Our hive design was introduced by Dr Giorgio Venturieri called rational hive from Brazil during an informal conversation (chat) on the internet. We tried to customize the boxes so they suit our environment and can resist water entering the hive by installing rebates on the edge of each layered box during the rainy season.

As per the management, we already alleviated the mortality rate and good methods in the division of hives are being practised. Also, proper sanitation is important to combat the infestation.

Soon, we would like to engage in the production of honey wine and other stingless bee products and encourage the local community to engage in stingless bee production.

20.2.1 Addendum: Grajo's Farm, Casiguran, Sorsogon.

On the way to Legazpi airport via Sorsogon City, we passed the Municipality of Casiguran and I remembered Grajo's Farm which I visited in this town a few years back. I decided to contact the owner, Leo Grajo, to ask him about the current status and if he would update me on the latest developments. This farm was featured in the Handbook of Meliponiculture Vol 1, published in 2016. Appended herewith is the update.

AHJ: Would you update your production rates and price range of your bee product packages? Colony multiplication to what numbers now. Any new source of feral colonies?

Leo: Sorry, I don't have data on the production rate. The price of stingless bee honey is P300 per 200 ml bottle. The dehydrated pollen is P150 per 75 grams

Around 250 colonies were shipped to different parts of the Philippines (100 went to the Norwegian Alliance Mission in Marinduque[28], 35 went to the Davao region, and the rest went to buyers within the locality and others.

Currently, we have a total of 400 colonies, 60 of which are feral colonies obtained from nearby barangay sites.

[28] It was from Leo Grajo that I got the contact of the coordinator of Agriculture Cooperatives in Marinduque. See Addendum 10.0

AHJ: How's the weather been? Any typhoon devastation?

Image 71 Every year typhoon brings devastation to our region

Leo: Every year typhoons bring devastation to our region. They bring an additional burden to farmers, especially coconut farmers. Typhoons devastate coconut plantations, which are the staple forage source of stingless bees, resulting in reduced production of honey, pollen and colonies.

The "hexagon meliponary" house is the only existing infrastructure that withstood the last storm. Additional more permanent stingless bee stands were constructed. These are typhoon-proof, maintenance-free, and easily accessible during splitting and harvesting. A stand consists of a concrete post and PVC pipes. A 20-foot stand can accommodate 8 to 10 colonies while a 10-foot stand can accommodate 5 to 6 colonies, respectively. The stands are set 15 meters apart from the next and are at strategic places in the farm boundaries to maximize the foraging of the bees.

In 2018, our farm suffered an infestation due to high density. Consequently, we constructed stands in 3 different areas: at our house in town, at the farm and the extension farm. Dispersing the bees in these separate locations reduces the infestation and results in stronger colonies due to less competition for forage sources.

AHJ: Do you have a transit area or sickbay of sorts?

Leo: Before shipping the bees either by land or by air freight we need to secure a certificate from the Bureau of Animal Industry (BAI) national office They will conduct an assessment of the farm and obtain samples from colonies for laboratory analysis, after the issuance of the certificate.

The certificate needs to be presented to the Provincial Veterinary Office for the issuance of a travel permit. The permit is valid for only 5 days.

The farm gate price of our colonies is P 3,500. The buyer picks up the colonies from the farm after attending an orientation training using Brazilian Technology hive design.

20.2.2 Developments[29] in Grajo's Farm

Image 72 Top left: The "Hexagon Meliponary" house is the only existing infrastructure that stood against a storm. Other photos: The farm gate price of our colonies is P 3,500 and will be picked up by the owner after orientation training using Brazilian technology hive design

[29] The signature Hexagon SB hive shack I highlighted in the "Handbook of Meliponiculture Vol1" has improved into the "Hexagon Meliponary" as Leo calls it.

20.3 Visiting Balay Buhay sa Uma, Bulusan, Sorsogon

Figure 313 A very impressive spread turned into a resort stay

Figure 314 Coco-shell hives and wooden box hives adorn this location in front of the chalets

20.3.1 Etymology: Balay Buhay sa Uma means "Home Life in the Farm" and is located in the Bulusan district which is named after the Volcano Mt. Bulusan.

Figure 315 The chalets are modern bamboo architecture excellent both inside and outside

20.3.2 Main Bee farm: Some 3,000 SB[30] (stingless bee; Roubik, 2021) colonies in coco-shell hives and wooden box hives adorn this location. It must have been a chore to put everything in place. It appears that the Landscape and Beescape have been well thought out as the hives are positioned about 2 meters apart. The flowering shrubs and feature trees like the Pili and the Avocado are strategically placed to provide foraging sources.

Images below show the surrounding natural vegetation and the varied cultivated vegetation which includes creepers, trees and shrubs in a well-designed landscape and the Beescape. Balay Buhay sa Uma Bee Farm is blessed with abundant groundwater and the owners utilize it well. The swimming pool is streamside at a higher ground and drains down to lower fishponds and well-designed water bodies.

[30] Editor's note: invented the term, along with HB, to make the word count shorter in my review paper, in Annual Review of Entomology, Vol. 68. "Stingless bee (Apidae: Apinae: Meliponini) Ecology".

20.3.3. Surrounding vegetation[31]

Figure 316 Surrounding vegetation with assorted flowering plants

20.3.4 Emergence of scouts: The first batch of scouts came out to stretch their wing muscles and warm up to prepare for forage recognisance at 5:30 a.m. Sorsogon time. Subsequently, the scouts (presumably)

[31] See also 20.3.5 Addendum – Avocado Flowers.

exercise their wings by 5:45 a.m., and they are at the avocado trees (see Addendum). The images below are from during the Civil Twilight.

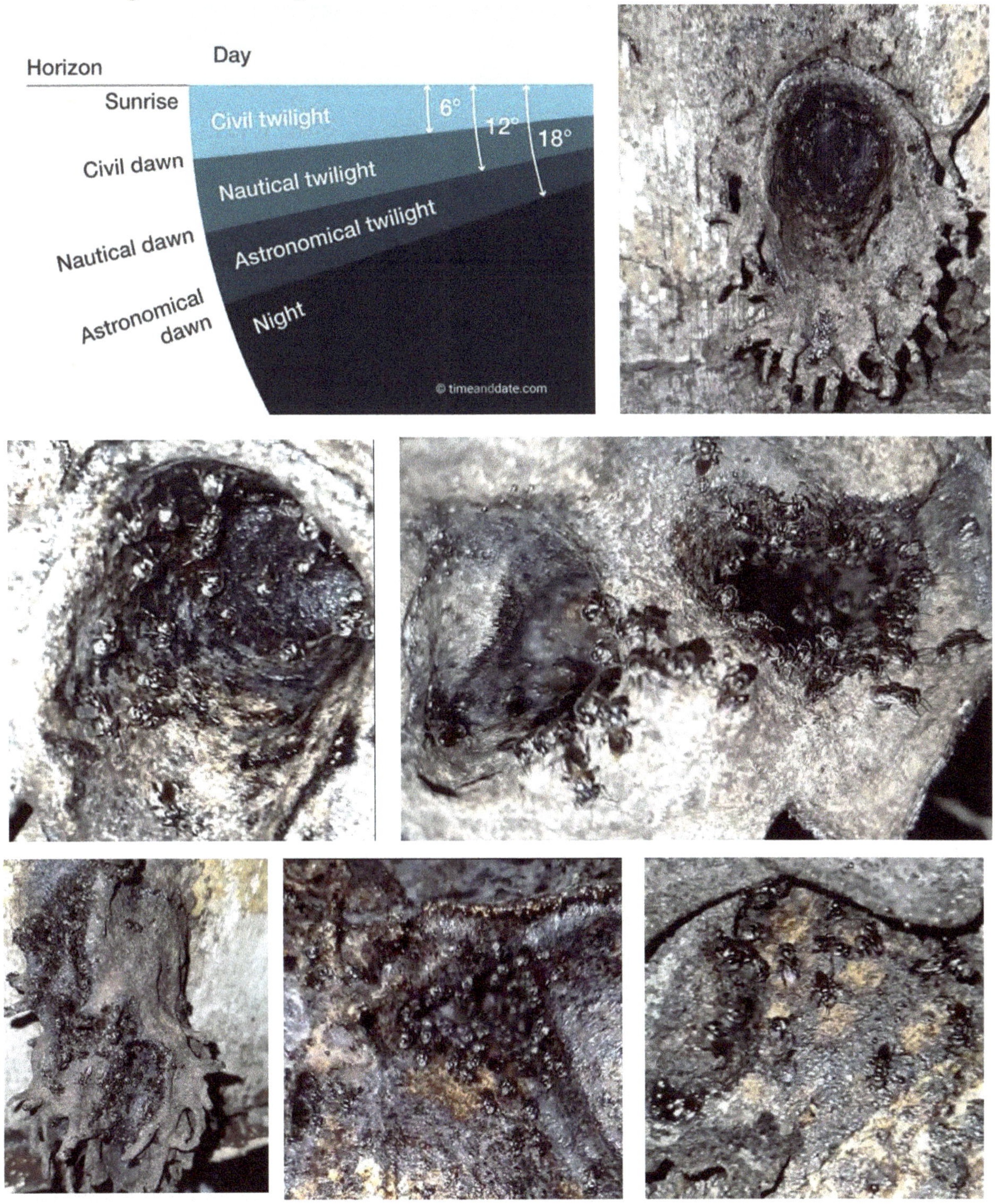

Figure 317 Emergence of scouts from domesticated hives in Balay Buhay sa Uma bee farm during civil twilight. The same timing occurs in both coco-shell hives and wooden box hives

20.3.5 Addendum – Avocado Flowers in Balay Buhay sa Uma

In the first week of March, the avocado flowering is in full bloom in Sorsogon, Philippines. Just the right time to be in Balay Buhay sa Uma to observe the flowering and pollination activities of stingless bees on avocado trees.

There is a cluster of about half a dozen trees behind the main kitchen of this bee farm "resort". The flowers switch from a female phase (when newly open and blooms in the time of day depending on ambient

Figure 318 A stingless bee on a male flower captured at 6 a.m. on the 2nd day at Balay Buhay sa Uma: right: A cluster of female flowers in the first week of bloom

temperature and humidity to a male phase the next morning. The **flower** has both female and male organs, but they don't function at the same time. Each **flower** is female when it first opens. That is, its stigma will receive pollen from other **avocado flowers**, but its stamens (male organs) do not shed pollen at this first opening. Trees must be the dwarf variety because they were all flowering at about 2.5 m to 4 m overall height.

Avocado flowers are set for about two to three weeks and, as already stated above, are dichogamous (meaning the stamens and pistils of a flower are 'activated' at different times) so that self-fertilization is prevented.

Figure 319 Foraging frenzy captured on the first day at Balay Buhay after lunchtime

20.3.6 Citation According to McGregor (1976), the nectar is secreted from the staminodes during the pistillate period and from the "true" nectaries during the staminate period. Sedgley (1985), on the other hand, reported that both staminodes and nectaries are active during the pistillate and the staminate stages. Reference: New Insight into Avocado Flowering in Relation to Its Pollination (Ish-Am and Eisikowitch, 1991)

20.3.7 Balay Buhay sa Uma Extension Bee Farm

Figure 320 Top row: SB colonies arranged in hanging rows similar as in the main farm; Middle row: Coconut plantation transformed into a stingless bee farm. Bottom row: buttress of Pili tree oozing resin and being foraged by the bees

BBsU extension farm, about 20 min. away is another large spread with much natural vegetation and coconut trees. Appears to me to be a coconut plantation being morphed into a bee farm. Ame Locsin shows a Pili Tree with its resin (elemi) oozing out at the buttress roots. The affinity for the bees to converge on that resin blotch has caused many to be drowned in the sticky blotch of resin.

Figure 321 Top Row: Coco-shell hives wrapped with fan palm leaves hung in rows. Middle left: Resting after some exhaustive surveying around. Middle right: Demo on hiving and colony splitting; Bottom left: a feral colony in a peculiar looking nest of ferns; Bottom right: Group photo upon completion of the farm tour

20.3.8 Hive splitting and feral colony extrication photos by DWR

Image 73 Photos by DWR: Colony splitting of coconut kernel shell hive. Bottom photo shows a climber being swarmed by thousands of bees while extricating a feral hive from the cluster of epiphytes.

List of Figures in Part 4

PART 5 - DAVAO REGION XI, MINDANAO

Mindanao (/mɪndəˈnaʊ/), still commonly known as Southern Philippines, is the second-largest island in the Philippines. Mindanao and the smaller islands surrounding it make up the island group of the same name.

Davao Region XI

Davao Region, formerly called Southern Mindanao (Cebuano: Habagatang Mindanao), is an administrative region in the Philippines, designated as Region XI. It is situated in the southeastern portion of Mindanao, comprising five provinces: Compostela Valley, Davao del Norte, Davao del Sur, Davao Oriental and Davao Occidental, the newly created province.

Bukidnon

Bukidnon, Cebuano for "people of the mountain," referring to the indigenous tribes inhabiting the Central Mindanao highlands. Early Spanish accounts give the name of these tribes as Buquidnones or Monteses de Mindanao ("mountain people of Mindanao"). Bukidnon eventually became the name applied to the territory they inhabited, which became a sub-province in 1907 and a province in 1914.

Cotabato (North and South)

Hispanicized form of Kuta watu, Maguindanaon for "stone fort," referring to an actual stone fort that stood on Tantawan (now Pedro Colina) Hill, around which grew the settlement that eventually became the capital of the undivided province. Subsequent divisions and the creation of new provinces have resulted in both North Cotabato and South Cotabato exercising jurisdiction far from their namesake city.

Davao (de Oro, del Norte, del Sur, Occidental, and Oriental)

Hispanicized form of the Bagobo word Dabo, meaning "to fall", alluding to the drop in elevation the river that now bears the name undergoes on its way from the slopes of Mount Apo to the sea. A related word in the neighbouring Obo language, davoh, means "beyond the high grounds" (i.e., the lowland). The settlement at the mouth of this river was also called Dabu; a 1628 document renders the native name in Dutch orthography as Daboe. The placename was later rendered in Spanish orthography as Dávao, and this settlement lent its name to the Spanish-era military district and, later, American-era province, of which it served as the capital.

Meliponiculture Distribution in Davao Region XI

Chart 11 Meliponiculture Distribution in Davao Region XI

List of Meliponiculture activities in Mindanao

1. Allan Bernante Damilag, Manolo Fortich, Bukidnon.
2. Boih Loyola, Loyola Farm, Davao Norte
3. Ace Lonzaga, Zambaonga, Mindanao
4. Mercy Namang Fuerzas, Tagum City Apiary, Davao Del Niorte
5. Peter Paul Rafosala, Panabo City, Davao Del Niorte
6. Ador Picaza, Oriental Bee Farm, Davao Oriental, Mindanao
7. Edwin Fetalino. Koronadal, South Cotabato
8. Oliver Nueva Ecija, Km5, Buhangin, Davao City
9. Gogo Manug, Hawilian, Agusan Del Sur
10. Cir Carlkendrick G. Tadlas, Malaybalay, Bukidnon
11. Glenn Estiva Lagumbay, Malaybalay, Bukidnon
12. Batoytoy Hermosa, Valencia, Bukidnon
13. Jake Hinay, Madaum, Tagum City

CHAPTER 21 - DAVAO REGION, MINDANAO

21.1 Oriental Bee farm, Lupon

I Visited Davao Oriental at the Oriental Beefarm in Lupon owned by Dr. Ador Picaza

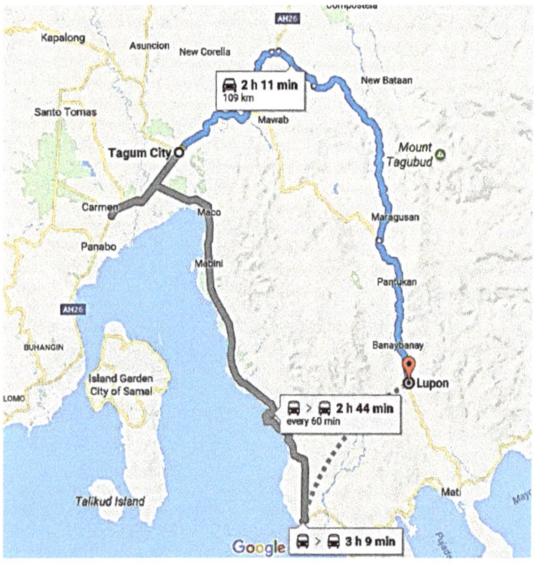

Map 6 Route to Oriental Beefarm in Lupon owned by Dr. Ador Picaza

Ground Orchids & Saluyot (*Corchorus olitorius*) is an edible leafy vegetable that is a member of the genus *Corchorus*, classified under the subfamily Grewioideae of the family Malvaceae. Found this at Oriental Beefarm in Davao Oriental, Mindanao, Philippines. It is frequented by *Tetragonula laeviceps* or a *T*. sp., shown to me by Dr. Ador Picaza, owner of Oriental Bee Farm. At his Apiculture Training centre, Dr. Picaza also keeps some *Tetragonula laeviceps* & *Tetragonula* spp. hives. Some are found in his wooden and bamboo furniture.

Figure 322 Ground Orchids

Saluyot (Corchorus olitorius)

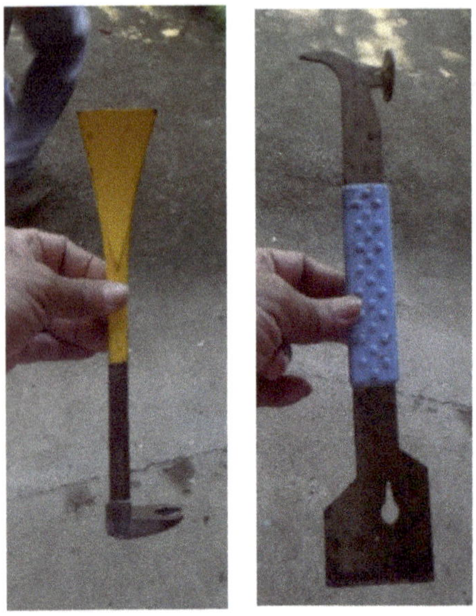

Figure 323 Apiculture Prying Tools

with Dr. Ador Picaza

Assorted Tetragonula sp. Box Hives

Figure 324 Some Tetragonula sp. Hives and nest entrances in the wooden

See *Figure 102,* we spotted a queen in one of the boxes and the bees were slightly larger than the others which I would only compare with *Tetragonula* cf. *dapitanensis*

Figure 325 Tetragonula cf. dapitanensis queen

Figure 327 Tetragonula biroi hives up to 50 units in this station

Subsequently, we visited his *Tetragonula biroi* farm on his coconut plantation. Peculiarly the almost 50 hives were placed closely and were reasonably healthy although they recently endured a typhoon.

All the hives were housed under an *Atrocarpus gomezianus* tree. The colonies obtained sufficient propolis from this tree. *Artocarpus gomezianus* (syn: *Artocarpus lakoocha*) provides resin and shade for about 50 box hive colonies on Oriental Bee Farm in Davao Oriental, Mindanao, Philippines... owner Dr. Ador Picaza

Figure 328 Artocarpus gomezianus (syn: Artocarpus lakoocha) provides sap and shade for about 50 box hive colonies

21.2 Loyola Bee farm

From Lupon Davao Oriental we then travelled to Panabo City in Davao del Norte

Figure 330 Loyola beefarm concentrates more on Apis melifera.

Figure 331 Visited Loyola Beefarm owned by Boih Loyola

Figure 329 Tetragonula sp. kept in an unconventional rectangular box hive.

Tetragonula sp is kept in an unconventional Rectangular box hive. Loyola bee farm concentrates more on *Apis melifera.*

We discussed feeding, where previously they used to feed sugar water in times of shortage of foraging sources, especially during extreme weather conditions like typhoons. The monsoons are rather different in the Philippines where the northeast monsoon called *Habagat* and the southwest monsoon called *Amihan* come back to back and there is a summer period in between these annual cycles. Anyway, nowadays Boih Loyola gets Jackfruit mush from a farm which he blends into a much-favoured feed among the bees.

21.3 Honey House Bees & Beeks

The next destination was the Honey House Bees & Beeks on Apokon Road in Tagum City.

Figure 332 Owners are Peter Paul Rafasola and his wife Claire Azuelo-Rafasola are former beekeepers in America. The box hive construction appears to have an Australian design influence with a clear plastic cover for observation. We encountered the Tetragonula hive in the woven bamboo wall

We encountered a *Tetragonula* hive in the woven bamboo wall. This farm also breeds *Apis mellifera* together with *Tetragonula biroi.*

The box hive construction appears to have an Australian design influence with a clear plastic cover for observation.

The owners are Peter Paul Rafasola and his wife Claire Azuelo-Rafasola are former beekeepers in America. They started in Tagum city in 2014 and specialize in bee pollen, comb honey, propolis and bees wax. They also offer hive equipment.

Figure 333 Meliponine & Apis beehives Honey House Bees & Beeks in Apokon Road in Tagum City

21.4 Tagum City Botanical Park Apiary

Finally, we were invited to the Apiary Botanical Park in Tagum City.

Figure 335 Unique Bee crossing sign; a healthy colony with much propolis

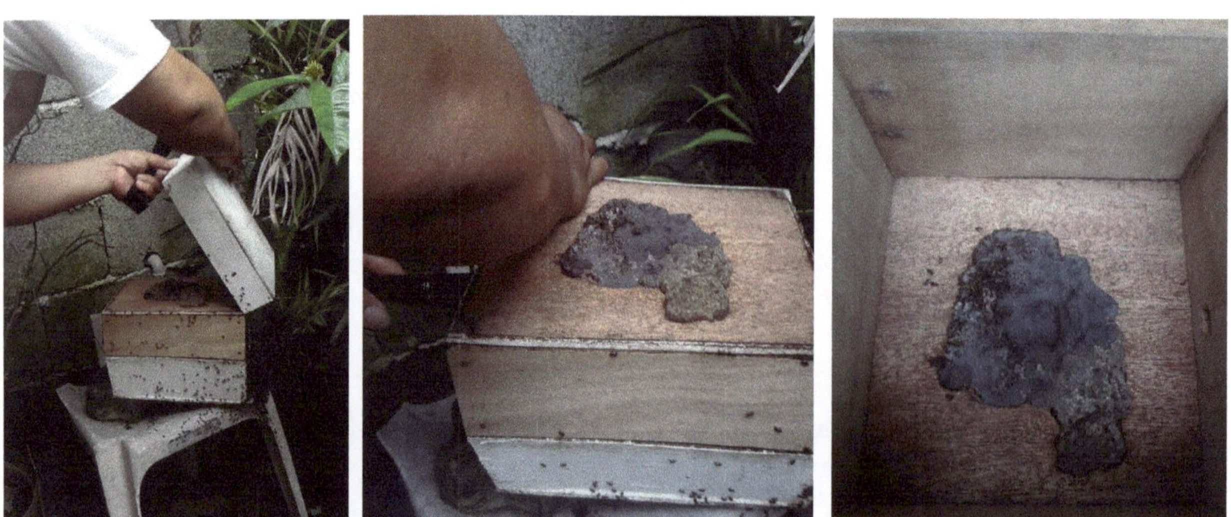

Figure 334 Weak colony with extension trials using Perspex material

The extension above the wooden hive is made of perspex and is very questionable as an insulating material. It was evident the colony did not proceed above the wooden material to expand its nest.

Figure 336 The extension above the wooden hive is made of perspex and is very questionable as an insulative material. It was evident the koloni did not proceed above the wooden material to expand its nest

Figure 337 With the Mdm Mercy Namang Fuerzas, custodian of the Botanical Park Meliponary & Apiary in Tagum City, Mindanao

21.4.1 Tagum City Fair

Figure 338Various Honey flavours

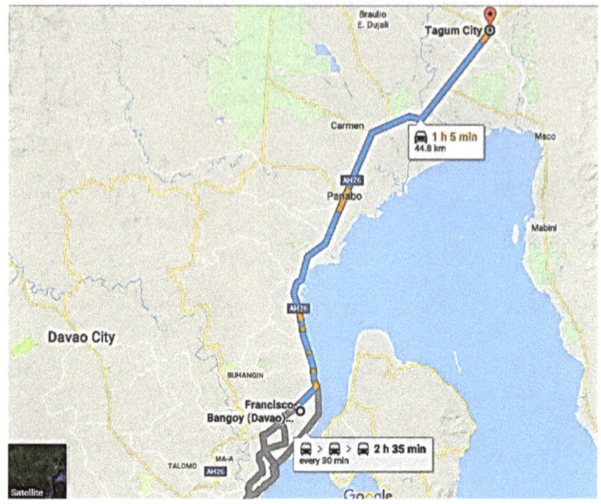

Map 7 Route to Tagum City

Figure 339 Assorted bee products and various kinds of honey.

I took a flight to Manila from Naga city to connect to Davao and was met at the airport by Peter Paul Rafasola and Michael Van Cogollodo. They brought me to Tagum city; coincidentally a Festival was going on. There were assorted products from the agriculture community. However, my interest was only in honey and bee products.

The Tagum Festival was a fun fair of sorts featuring local and regional products. I was brought to the Honey products stall displaying and selling beekeeping products from a cooperative group in Davao.

CHAPTER 22 – OTHER REGIONS IN MINDANAO

22.1 Edwin Fetalino @ Sultan Kudarat, Mindanao

Feta Integrated Farm Technical Institute Inc., is located at South Sepaka, Bagumbayan, Sultan Kudarat, Mindanao, Philippines. ATI 12 (Agricultural Training Institute Region 12) learning site for agriculture, a school for practical agriculture, SOX'S (SOCSKSARGEN) farm tourism member, 3rd party certified organic farm by OCCP (organic certification centre of the Philippines). At present, it has 148 colonies (133 colonies in Embrapa boxes and 15 colonies in coco hives) both *T. biroi* and *T. laeviceps*. We practised urban beekeeping, farm beekeeping, and agro-forestry beekeeping. We produced honey for personal consumption and bee pollen in chilli garlic sauce. We sell our colonies through the Department of Agriculture Dispersal of Organic Beekeeping project.

CERTIFICATE

Certificate No. 053720-2

FETA INTEGRATED FARM

CERTIFIED OCCP TRUE COPY

Producer Code: 19070185

Edwin H. Fetalino, Farm Owner
Sitio South Sepaka, Bagumbayan, Sultan Kudarat, 9810

The following is in compliance with PNS 07:2016 and OCCP ICSI certification Requirements

Product/Product Category	Brand/Trade Name	Area	Units	Quantity	Status
Chicken Eggs	NA-No labelling done	1.50 hectares	1 FARM	14,600 PIECES	Organic
Stingless Bee Colony				60 COLONIES	Organic

Annex(es): APPROVED FARMERS AND PRODUCTS LIST

This certificate covers the following Activities: Animal Production with Beekeeping and Trading
The certified operator also produces/processes non-certified products: YES Refer to APL

LEILANI RAMONA K. LIMPIN
Chief Executive Officer

Date Certification Granted: 22-December 2020
Expiration Date: 21-December 2021

Applicable seal(s):

19070185

Organic PHILIPPINE DEPARTMENT OF AGRICULTURE

OAA-OCC-N2018-001

Note: To issue this certificate, corrective actions have to be in place to address the identified non-compliances. The effective duration of this certification depends on the immediate compliance of the operator to this.

The operation has been inspected during the production year by an accredited OCCP inspector to [illegible] best of our knowledge, the standard have been met. This only refers to [illegible] production and not to any other aspect of food quality

Image 74 3rd party certificate for organic beekeeping

3rd party certificate for organic beekeeping

Image 75 Edwin H. Fetalino, owner of the farm & precious citadel, tris, my wife Emily and Shane

Edwin H. Fetalino, owner of the farm, email: edwinfarm2020@gmail with contact number 09057152285 and his family (precious Citadel, Tris, my wife Emily and Shane) involved in beekeeping

Image 76 Farm entrance with different flowering plants, source of nectar and pollen for stingless bee Tetragonula biroi at vertical hive (Embrapa box)

Tetragonula biroi showing the distribution of brood, honey and pollen. Originally the Embrapa box[32] was developed by Giorgio Venturieri of Brazil, and adapted by Leo Grajo of Sorsogon (Grajo's Agri-bee

Image 78 Tetragonula biroi showing the distribution of brood, honey and pollen.

farm), the box dimension is 3inch height, 1inch thickness and 8inch width, the box is composed of a brood chamber and honey chamber, with bottom and top cover Embrapa (Empresa Brasileira de Pesquisas Agropecurias)

Image 77 Coco hive showing brood chamber

[32] EMBRAPA, Empresa Brasiliera de Pesquizas Agropecuarias

22.2 Stingless beekeeping in Mindanao, the Philippines; by Ace Lonzaga

History

I got interested in beekeeping last 2013. Initially, I was planning to keep *Apis mellifera* but was discouraged by its acquisition and maintenance costs. Aside from that, imported bees are prone to diseases that would require additional costs for a chemical cure.

So I started to look for other alternatives like *Apis cerana* which is common and native to the Philippine forests. However, I've read that only a few *Apis cerana* beekeepers are successful in keeping them due to their absconding behaviour.

While reading about beekeeping in the Philippines I was surprised to know that there are stingless bees that are indigenous to the Philippines. It was the first time I heard about the stingless bee and was curious about it.

I began researching these bees and found how different they were from the *Apis* species. At that time I only knew about *Apis* and have always thought that all bees have stingers and hexagonal-shaped brood cells.

22.2.1 Meliponiculture

My interest in stingless bees grew each day and I started to look for beekeepers who sell colonies. One species being promoted for large-scale pollination and production of honey, pollen and propolis by UPLB (University of the Philippines Los Baños) and most stingless beekeepers is *Tetragonula biroi*. Unfortunately, the initial information was that this species is only found in Luzon (northern Philippines) while I am situated in Mindanao (southern Philippines). However, I was determined to have a couple of colonies. I purchased them and they were delivered by aeroplane cargo. (*T. biroi is also present in Visayas and Mindanao, based on biological records and personal observations. We collected samples of T. biroi in Guimaras Island, Bacolod, Iloilo and Panabo, Davao city. pers. comm. Cervancia, 2016)*

Upon seeing the actual bees I was amazed because they looked so familiar. I then began to search for that very familiar flying insect in our backyard. Lo and behold, I found its nest in one of our junk storage bins. I was very happy to know that there are stingless bees in our area. I then prepared a box hive so I could transfer it instantly.

Upon checking the colony, there was one thing very different from it based on what I've read about *Tetragonula biroi*. Its brood pattern was different. There was no spiral shape common to *Tetragonula biroi*, instead, it has a clustered pattern. Upon further research, there are species which have a clustered pattern. Sharing what I've discovered on social networking sites and forums I learned that it was called *Tetragonula laeviceps*.

Tetragonula biroi

One species that can be found throughout the Philippines is *Tetragonula biroi*. It is said to be only seen in Albay and some other parts of Luzon. (*T. biroi* is widely distributed in Luzon, but the populations are high in the Bicol region, Quezon, Laguna and Batangas. pers. comm. Cervancia, 2016). The Bureau of Agricultural Research is calling these bees the bees of the future (source: http://www.bar.gov.ph/digest-home/digest-archives/369-2013-1st-quarter/4741-janmar2013-stingless-bees).

One of *Tetragonula biroi*'s distinctive traits is its spiral brood pattern concentrated in the middle of the hive. Not all colonies exhibit this trait as I've seen some with irregularly positioned brood comb.

Another thing you've probably noticed with this species is how they gather around their hive entrance. (A common behaviour among social bees. pers. comm. Cervancia, 2016) Before I thought they were all guard bees performing their duties but I was told by an experienced *T. biroi* beekeeper from Albay that most of the bees are processing the gathered nectar. They are doing it outside where the temperature is higher to reduce the moisture content, turning it into honey (although this has no scientific verification; pers. comm. Cervancia, 2016). Also, they create an entrance tube made from a sticky substance called propolis which will become their flight and landing path. Guard bees are aggressively monitoring this part of the hive for intruders. Come extremely close and you will know what I'm talking about.

To have an idea of how productive these bees are, below is a photo of six months of hard work by the bees. Of course, this is not the same all year round as the season changes. The best time for the bees to produce is during the honey flow season (around summer) when the trees start to bear fruit.

Unlike honeybees, stingless bees don't build honeycombs for honey storage. Instead, they build honeypots and pollen pots. This is a unique trait that probably an average Joe has never heard of. The pots are made from a mixture of tree resin, wax and a special substance from the bees. The result is an anti-bacterial substance that is made into pots and preserves the honey which can last many years without spoiling.

Tetragonula laeviceps

Another stingless bee species found in the Philippines is *Tetragonula laeviceps*. (see *Figure 115*Figure 115 e to k). It is very similar in size to *Tetragonula biroi* and looks almost the same to the naked eye. You will only be able to see the difference under the microscope or a macro lens.

This bee is commonly found in Mindanao, the Visayas and some other parts of Luzon. Compared to the Biroi species, the number of bees in a colony is much less and they are not that aggressive. Colonies may reach an estimated size of only 1000-5000 adult bees while the Biroi could double or triple that number depending on the availability of resources.

One of the major differences between the Biroi and the Laeviceps is the brood pattern which can be best described as a mesh network of brood cells. (see *Figure 115*). The brood varies in colour from dark to light brown which indicates maturity. Dark brown cells are immature cells while light brown is almost ready to hatch.

As with any stingless bee, Laeviceps also builds a flight and landing path in an entrance made of propolis. The difference, however, is that they intentionally cover it with moss-like objects for which I still don't have any explanation. Also, they only make a very small opening in the entrance which is heavily guarded by the bees. If one looks closely at the entrance opening, all you can see are heads sticking out ready to defend the hive. If a worker bee approaches or leaves the hive, the guard bees give way and are doing it flawlessly as if they talk to each other considering the number of bees flying in and out.

Another distinct trait of this species is that a guard or worker bee hovers in front of the entrance as if surveying the entrance. I've noticed bees hover for hours and enter the hive after some time. The number of bees hovering ranges from one to three. What I am uncertain about though is if the bees come from the same hive or are spies from another colony.

The Laeviceps is not a very prolific honey producer and is less active in foraging, probably because of the limited number of bees. It is best not to harvest and just leave the honey for the bees to consume. Like other stingless bee species, they make honey and pollen pots for storage which are just amazing to look at.

Bait hive experience

Honey flow season is not only the best time for harvesting honey but also for propagating bee colonies. Bees will most likely survive a split during these months due to the abundance of the food supply.

However, I don't want to directly split my colonies as I've had several bad experiences before. Other more natural and less harmful methods are available like eduction and bait hives.

I have been closely monitoring my colonies for a couple of months (April-May) waiting anxiously for their box to be full of honey and pollen. As soon as the bees filled their boxes, I started to set up bait hives around our house where the bees could colonise.

I used the boxes I made several months ago and applied beeswax and propolis inside. I've read on the internet that this will attract the bees and encourage them to settle in. For that reason, I placed around seven boxes outside about ten meters away from my stingless bee colonies. As weeks passed no bees were seen trying to inspect my boxes. Several of my bait hives were already occupied with different insects like lizards, spiders, ants, etc. Weeks turned to a month and still, nothing happened. So I thought, the bees are not ready to propagate yet because of the hot weather we're experiencing here in the Philippines due to El Niño.

I decided to just remove my bait hives and try it some other time. An old furniture shelf with drawers was taken out from my parent's bedroom and put outside the house to rot. The condition of the furniture wasn't that bad so I made it my tool shelf. I moved in some of my tools, beekeeping equipment and bait hives. I officially ended my baiting attempt since it has already started raining daily here in our place. I thought the honey flow season has ended and the bees are in no way possibly interested to look for other homes now.

Image 79 a) Apis sp. b) box hives c) Tetragonula biroi cluster brood cells d) T. biroi worker photo: Texter Taclobao e) Box hive nest entrance of T. biroi f) & g) Tetragonula laeviceps brood h) T. laeviceps nest entrance

22.2.2 **Efficacy**

Then something great happened! I began noticing a swarm of stingless bees flying close to the old furniture where my bait hives are stored. I initially thought it was a bunch of mosquitoes as the weather was not that good at that time for bees to be flying out. But to my surprise, it was a swarm of stingless bees (probably *Tetragonula laeviceps*). So I immediately inspected which of my bait hives they have started to build their new home in. Interestingly, none of my bait hives was occupied. I started to inspect other parts of the furniture and saw that they have settled in one of the drawers.

I felt confused and disappointed at that time. Why would the bees settle in a rotten and smelly drawer and not on my bait hive with beeswax and propolis? So I removed the drawer and saw they were already building honey and pollen storage in preparation before the queen's arrival. I placed one of my bait hives in the position of the drawer thinking the bees would be fooled to settle in the hive and not the drawer. But hours passed and the bees began to disappear. So I returned the old drawer and the bees started to come again.

I guess the bees had decided to settle into the drawer and nothing could be done to persuade them otherwise. The swarm had started to grow and more bees were flying in and out of the old drawer. I have observed many of them bringing resin and pollen, on their hind legs, inside. I think I will have to settle for this unexpected but good result.

I'm not exactly sure how I can advise beekeepers to successfully bait a swarm of stingless bees. Perhaps, don't waste your time building bait hives because bees prefer furniture. But really, you could only hope that bees would choose to settle in your bait hives. Bees know best and when they have chosen a spot you can do nothing. Just sit, relax and enjoy nature at its best!

22.2.3 Studies in *Tetragonula biroi* flight patterns

Take-off and landing strategies of Indo-Malayan Stingless Bees

(*Tetragonula biroi* Friese)

(Windy Day); by Ace Lonzaga

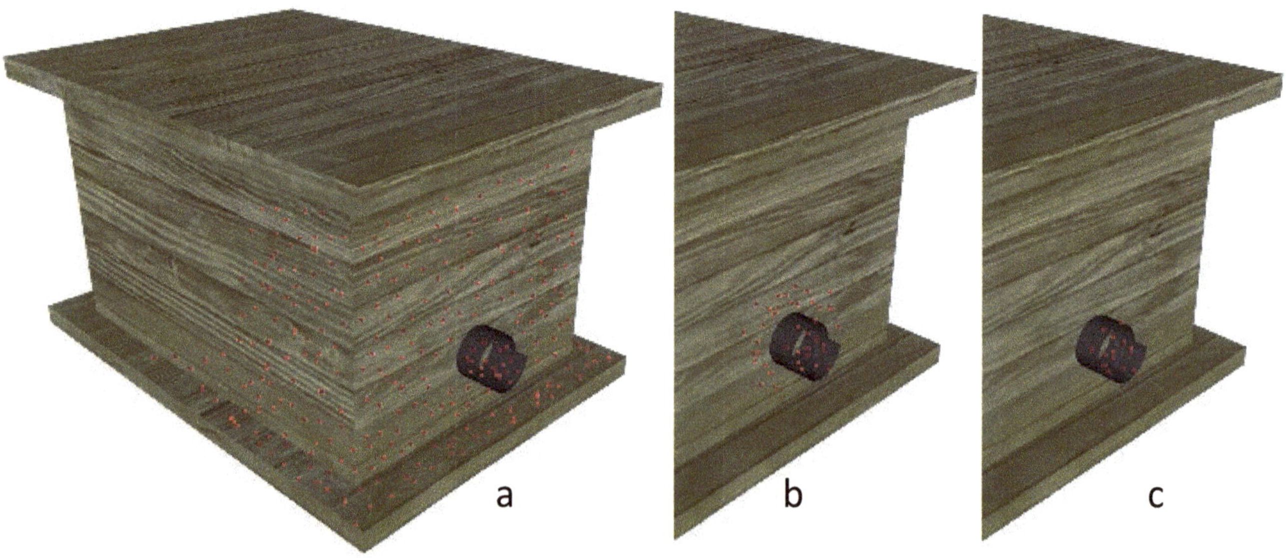

Image 80 a) Some bees will occupy the sides of the wooden box hive b) Guard bees are assigned to guard the tube c) 95% of bees will take off from the lip/brim of the tube

1. Bees are represented by red dots.

- Early morning, bees only occupy the tube entrance
- From 8:00 am to 10:00 a.m., bees will gradually increase in number and will start occupying the front panel of a wooden box hive. Depending on the weather, the number of bees that are present outside will vary. On a gloomy day, the number of bees is minimal. However, a sunny day with the wind will still encourage bees to go out. Probably, this happens because they have gathered more nectar which they will need to process outside their hive.

- There are approximately 800-1000 guard bees at their peak (12:00-5:00 p.m. assuming a sunny and windy day). Some will occupy the sides of the wooden box hive. (see *Figure 117* a)

- Bees will start getting inside the box as the sun sets (5-6:30 p.m.). Guard bees are assigned to guard the tube (see *Figure 67*b).

2. Take-off pattern

- Almost 95% of the time the bees will take off from the lips/brim of the tube as shown (Figure 116 c & Figure 117a) (same on a regular sunny, calm day).

It is only when there is too much traffic that the bees will try to crawl under/outside the tube and take off from there (Figure 117 c) (same on a sunny and calm day).

3. Landing Pattern

- Upon landing only about 60% of the bees can land inside the tube. (*Figure 67* b)
- The other 40% will randomly land on different parts of the box close to the entrance. Most can land on the external part of the tube and just crawl their way in. The wind has a significant effect on the percentage of their successful landing (Figure 117 d).

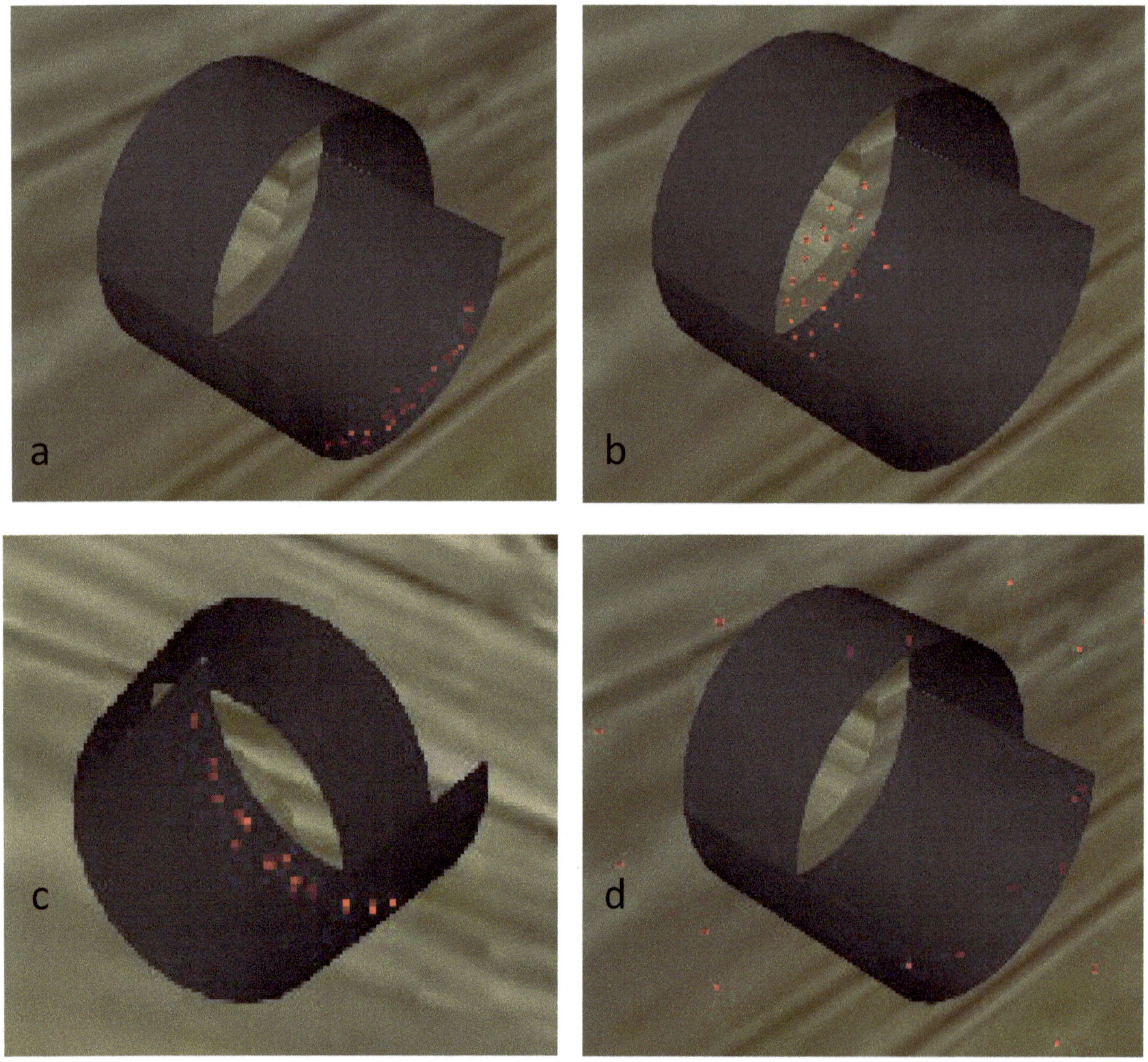

Image 81 a) Bees take off from the edge of the tube; b) 60% of bees can land on the inside of the tube; c) bees will try to crawl under/outside the tube during high traffic; d) 40% will randomly land on different parts of the box close to the entrance

4. Flight pattern when landing and take-off

    - The zigzag flight pattern that the bees show on a regular sunny and calm day is affected by wind disturbance. Some bees will no longer zigzag but fly directly straight-up to about 30 ft (same on sunny or calm days). Also, you can see other bees still doing the zigzag pattern but wider in dimension, as seen in the image below (*Figure 124*).

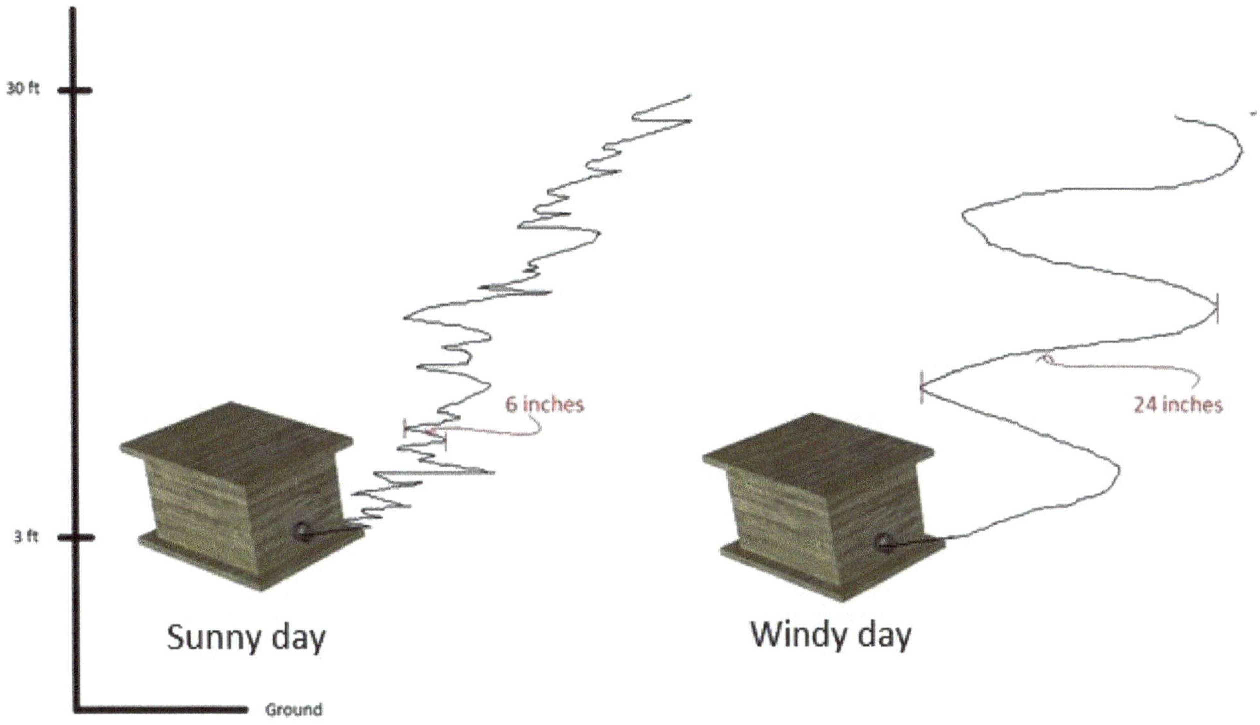

Image 82 The zigzag flight pattern which the bees show on the regular sunny on-windy day gets affected by wind disturbance

The zigzag flight pattern that the bees show on regular sunny and calm days is affected by wind

- The landing pattern also reflects wind disturbance. More bees no longer display zigzag flights or hover when winds are brisk. They seem to glide straight into the hole at a much higher speed. The landing success percentage, however, is less than on a relatively sunny and calm day.

References

Anonymous. (2006). Raising stingless bees (*Trigona biroi* Friese) using Japanese-designed hives. University of the Philippines at Los Baños.

Ciar, R. R., Bonto, L. S., Bayer, M. P., Rabajante, J. F., Lubag, S. P., Fajardo, C. A., & Cervancia, R. C. (t.t). *Foraging Behavior Of Stingless Bees (Tetragonula Biroi Friese): Distance, Direction And Height Of Preferred Food Source.*

Cockerell, T. (1919). Social Bees of Philippines. *The Philippine Journal of Science.*

Heard, T. A., & Hendrikz, J. K. (1993). Factors Influencing Flight Activity of Colonies of the Stingless Bee *Trigona carbonaria* (Hymenoptera, Apidae). *Australian Journal of Zoology · January 1993.* doi:10.1071/ZO9930343

Hilário, S., Imperatriz-Fonseca, V., & Kleinert, A. (2000). Flight activity and colony strength in the stingless bee *Melipona bicolor bicolor* (Apidae, meliponinae). *Revista Brasileira de Biologia, 60 (2)*, 299-306.

Sommeijer, M. J., Bruijn, L. L., Meeuwsen, F. J., & Slaa, E. J. (2003). Reproductive behaviour of stingless bees: nest departures of non-accepted gynes and nuptial flights in *Melipona favosa* (Hymenoptera: Apidae, Meliponini). *Entomologische Berichten , 63*(1), 7-13.

Smith, J. P., Heard, T. A., Beekman, M., & Gloag, R. (2016). Flight range of the Australian stingless bee *Tetragonula carbonaria* (Hymenoptera: *Apidae*). *Austral Entomology · January 2016.*

Sommeijer, M. J., Bruijn, L. L., Meeuwsen, F. J., & Slaa, E. J. (2003). Reproductive behaviour of stingless bees: nest departures of non-accepted gynes and nuptial flights in Melipona favosa (Hymenoptera: Apidae, Meliponini). *Entomologische Berichten , 63*(1), 7-13.

Sung, I. -H.; Yamane, Soichi; Lu, S. -S.; Ho, K. -K. (2011). Climatological Influences on the Flight Activity of Stingless Bees (*Lepidotrigona hoozana*) and Honeybees (*Apis cerana*) in Taiwan (*Hymenoptera, Apidae*). *Sociobiology 58(3) January 2011*, 835-850.

Temaru, E. (2007). Antibacterial Activity of Honey from Stingless Honeybees. *Polish Journal of Microbiology*. Retrieved from www.pjm.microbiology.pl/archive/vol5642007281.pdf

List of Figures in Part 5

PART 6 - APPENDICES

T. cf *dapitanensis* (Davao Or.) | *T. sapiens* (Iloilo) | *T. clypesris* (Aust)

T. biroi (Bicol) | *Tetragonula. sp.* (Takpal Biliran) | *T.* cf *sapiens* (Isabel)

T. laeviceps (M'sia) | *T. fuscobalteata* (M'sia) | *T. biroi* (Mindanao)

Figure 340 Tetragonula spp. Queens available in the Visayas

Distribution of Stingless Bees in The Philippines

Table 2 Distribution of Stingless Bees in The Philippines

Year	Author	Name	Synonym or alt. comb.	References	Distribution
1854	Smith	*Trigona iridipennis*	*Tetragonula iridipennis*	Balthazar 1966 p.407	Mindanao: Lanao. Zamboanga; Palawan
1857	Smith	*Trigona laeviceps*	*Tetragonula laeviceps*	Rasmussen 2008	Philippines Schulz 1907
1857	Smith	*Trigona nitidiventris*	*Lepidotrigona nitidiventris*	Balthazar 1966 p.407	Palawan?
1857	Smith	*Melipona laevicep?*		Balthazar 1966 p.407	Listed Philippines Brown (1906)
1898	Friese	*Trigona biroi*	*Tetragonula biroi*	Cockerell 1919 p.77	Luzon, Los banos & Bataan, Lamao
1908	Cameron	*Trigona fuscobalteata*	*Tetragonula fuscobalteata*	Starr & Sakagami 1987	Dumaguete & Valencia, Negros Oriental
1909	Friese	*Trigona luteiventris*	*Heterotrigona erythrogastra*	Cockerell 1919 p.77	Palawan, Puerto Princesa
1909	Friese	*Trigona laeviceps* var. *clypearis*	*Tetragonula clypearis*	Rasmussen 2008	Philippines Dollin 1997
1909	Friese	Trigona (Heterotrigona) clypearis	*Tetragonula clypearis*	Dollin et al. 1997	Philippines Dollin 1997
1911	Cockerell	*Trigona sapiens*	*Tetragonula sapiens*	Starr & Sakagami 1987	Baybay & Inopacan, Leyte; Negros Is.
1915	Cockerell	*Trigona palavanica*	*Lepidotrigona palavanica*	Cockerell 1919 p.77	Palawan, Puerto Princesa
1918	Cockerell	*Trigona valdezi*	*Tetragonula valdezi*	Balthazar 1966 p.407	Luzon: Laguna; Mindanao: Davao, Lanao.
1925	Cockerell	*Trigona dapitanensis*	*Tetragonula dapitanensis*	Cockerell 1925a: 492:	Dapitan, Mindanao

APPENDIX 1

Beekeepers Network Philippines Foundation Inc. (Beenet Philippines)

Background:

Beenet Philippines is a 26-year-old organization registered with the Securities and Exchange Commission. It has nine Board of Trustees headed by Epifanio Loyola in Mindanao. It has three BOT in Visayas headed by Alberto V. Bartolata. Beenet Visayas has been conducting several conferences in Cebu and co-hosted by the Department of Agriculture Region 7, in collaboration with BEES Association, BANAAG, CHAI, and PhilBee Hives. The conferences have provided a forum for stakeholders in the regional beekeeping community to address current challenges and opportunities facing the industry.

The output of the 1st congress is crafting the road map for Central Visayas and the 2nd congress is the drafting of the road map in the three regions of Visayas. Beenet has Guests from DA-BAR, DA-RFO7, DENR7, DOST7, DTI7, PCA and DBP for funding support. The road map for Central Visayas is presented below. The road map for the Visayas is in the PowerPoint presentation.

I. EXECUTIVE SUMMARY

1. The Central Visayas Beekeeping Road Map for 2020-2015 is the road map for policymakers and stakeholders of beekeeping in Region 7. It was crafted through the initiative of the Department of Agriculture, Regional Field Office 7 under the Enterprise Development Through The Establishment of Sustainable Livestock and Poultry Multiplier Farms Program of D.A. Beekeeping is included under this program implemented by the Livestock Section of Regional Field Offices all over the country.

The crafting of the Road Map involved the participation of beekeepers, academic institutions, the private sector, Non-Government Organizations, and government agencies in a series of workshops, consultations, and strategic planning activities.

2. The Beekeeping Road Map of Central Visayas shall be implemented and monitored by the Steering Committee created in partnership with the Department of Agriculture RFO 7 as the lead agency. Under the Steering Committee are the different committees designated to implement and monitor various activities for the development of the industry.

3. The status of Beekeeping in Central Visayas was reviewed, and constraints on its growth and facilitative factors in the attainment of its objectives were identified.

4. Strategies to address the high demand for bee products, biosecurity measures, continuous supply of queens and bee colonies, financing, multidisciplinary research programs, training and capacity building and conservation and management of indigenous bee species were given emphasis.

II. INDUSTRY PROFILE

Honey production is the primary bee product produced by beekeepers in Central Visayas. The average annual yield per colony is 40 kilograms. Other products include pollen, beeswax, propolis, selling of queen bees, colonies, tools, and equipment.

Purple Bee Shop and Café on Plaridel Street in Mandaue City, Cebu sell value-added products like mead or honey wine, honey and propolis soap, massage beeswax candle, facial cream, ointments, etc. They also served as a training centre for beekeepers.

Bohol Bee Farm in Dauis, Bohol also sold honeybee products with branches in the City of Tagbilaran and Cebu City.

Generally, honey production in most areas of Central Visayas may start in February and peaks from April to June. A dearth period (few flowers in bloom) usually starts from July to December. The majority (about 80 %) of bee products produced were taken from domesticated European honeybees (*Apis mellifera*).

Other bee products harvested were from indigenous bee species *Apis dorsata* ("Pukyutan"), *Apis cerana* ("Ligwan"), and stingless bee *Tetragonula* ("Kyut"). About 80 % of beekeepers were found in the province of Cebu.

Technology for mass production of stingless bees was beginning to recognize particularly in the province of Bohol, Cebu, and Negros Oriental. They were appreciated more for their pollination activities and improving crop yield. Honey and propolis from stingless bees were sought by consumers and traders being more organic and superior for their medicinal properties.

Most beekeepers market their products directly to the consumer. They seldom encountered marketing problems with their bee products produced.

Research activities in beekeeping were minimal conducted by the Research Division on D.A RFO 7, Bohol Island State University in Bohol and University of San Carlos Biology Department and Physics in Talamban Campus, Cebu City.

Three Beekeeping Associations existed only in the province of Cebu. Cebuano Honeybee keepers Asso. Inc. (CHAI) existing since 1993, Bees for Environment & Economic Sustainability (BEES Assn.) is a five years old organization and was recently created in 2015, the Balamban Apiarist for Nature and Agriculture (BANAAG) in Balamban, Cebu. Members of this association raised the European bee species while some of them simultaneously raised the stingless bee.

Other associations actively involved in beekeeping were found in Bohol namely Kapunungan sa Bol-anong Organikong Mag-uuma (KABOOM), Permakultura Bohol, and Gawad Kalinga. All of them raised stingless bees for bee products and sustainable pollination of their crops.

Other isolated beekeepers were not members of any organization spread throughout the four provinces of Region 7.

III. KEY CHALLENGES OF BEEKEEPING IN CENTRAL VISAYAS

An output was conducted by beekeepers and other stakeholders in Region 7 during a consultative beekeeping workshop.

A. STRENGTHS

Internal Facilitating Factors

1. Presence of many skilled beekeepers
2. Presence of three strong beekeeping associations, three farmer associations, and private and state universities with their beekeeping projects/programs
3. Increasing interest in beekeeping
4. Availability of beekeeping supplies and equipment
5. Adaptable native bee species
6. Currently, no report of fatal pests and diseases

External Facilitative Factors

1. High demand for bee products in the local and foreign market

2. Presence of research/laboratory institutions in D.A RFO 7, University of San Carlos, and Bohol Island State University supporting the development of beekeeping
3. Expanding need for pollinators in crops managed by commercial farms
4. Possible inclusion in the annual budget for beekeeping under the D.A Livestock and Poultry Program, High-Value Commercial Crops, and training activities of Agricultural Training Institute (ATI)

B. **WEAKNESSES**

Internal Hindering Factors

1. Inadequate supply of queen bees (*Apis mellifera*)
2. No regular communication/meetings among beekeepers, beekeepers' associations, and other institutions involved in beekeeping
3. Unscrupulous beekeepers selling weak, substandard nucleus colonies, and old/inbred queens.
4. Migration of skilled beekeepers to other countries for beekeeping work that offers greater income and opportunities

External Hindering Factors

1. No standard pricing for honey and other bee products
2. No trained quarantine officer for regulatory control of fatal diseases like American Foul Brood Disease, Hive Beetle, and contaminated beekeeping supplies
3. Lack of support from the government when it comes to financing, capability building, regulatory control, research and extension activities for beekeepers
4. Climate change and Free Trading

C. **OPPORTUNITIES**

1. Livelihood (Income Opportunities)

- Honey, pollen, beeswax, bee colonies, royal jelly, and propolis are accepted food supplements and often have medicinal properties in healing many ailments and are relatively in high demand in the local market and abroad.
- Value-added products from bee products can be made into honey wine, soap, candles, cosmetics, honey cider, pastries, cakes, and food supplement for game fowls, and are widely used by pharmaceutical companies.

- Selling bee colonies, beekeeping supplies, tools, and equipment to new and old beekeepers

2. Generates Employment

- For beekeepers, processors, retailers and carpenters in making beehives, equipment, etc.
- Many Filipino beekeepers were employed in Australia, Canada, New Zealand, the Middle East, and the USA working as beekeepers.

3. Increases Agricultural Productivity

Bee pollination enhances agricultural productivity and hence addresses the food security program of the government. Below are tables showing a yield increase of some crops pollinated by bees.

4. Enhances Biodiversity

Bees are responsible for pollinating crops, fruits, forest trees, wild plants, and mangroves. They are needed for the good seed formation of many plants. As Albert Einstein opined, "If the bee disappears from the surface of the earth, man would have no more than four years to live".

D. **THREATS**

A. Production

1. The decreasing number of beekeepers, bee colonies, and production of bee products due to the difficulty and unavailability of queens (European bees) over the past years

2. Decreasing the number of skills that raise quality island-born queens; many of these skilled beekeepers work abroad on commercial bee farms offering bigger salaries and privileges for them and their families

3. Inadequate support from a government agency (D.A, ATI, DENR, LGUs) to train new beekeepers in basic beekeepers in beekeeping and upgrade/capacitate existing beekeepers in advance beekeeping and queen-rearing seminars

4. Lack of implementation of Best Management Practices in Beekeeping by concern agency governing standards for bee colonies, Phil. Honey and other products

5. Minimal intervention in the promotion of native bees and their commercialization; our native species are more adapted to our local conditions and need less added input, and therefore, are more sustainable for small farmer livelihood

6. The proliferation of fake and adulterated honey.

B. Biosecurity

1. Lack of trained personnel in the quarantine station of the Bureau of Animal Industry for possible detection of pests and diseases during the transport of bees in and out of region 7

2. Imported queens and supplies that don't pass quarantine may transfer fatal bee pests and diseases in the region and the whole country

C. Research, Development and Extension

1. Lack of research where the weaknesses of the industry were identified; threats that may hamper the growth of the industry should be given immediate attention.
2. Inadequate funding support to implement needed beekeeping research proposals
3. Research institutions like D.A. and the academe don't have a regular annual beekeeping budget for research, development, and extension programs.

IV. FUTURE DIRECTION (2020-2025)

A. Vision - A profitable and sustainable beekeeping industry in Central Visayas driven by productive and progressive beekeepers and stakeholders increasing and conserving the bee population in the region

B. Mission - To help and empower the beekeeping stakeholders to produce enough, quality beekeeping products increasing their farm income while conserving biodiversity in Central Visayas

Road Map Goals and Objectives

Target	Output	Outcome
A. To increase the production of bee colonies, quality honey and other bee products	1. To produce quality queens (mellifera) for the replacement and expansion of bee colonies in the region 2. To buy quality local/imported queens to improve the genetic pool and distribute to capable beekeepers/beekeeping assoc./LGUs/ and academic institutions	• Many queens produced • Local and imported queens acquired and distributed • New beekeepers trained ➢ Training on sustainable queen rearing, wax pollen,

	3. To train more beekeepers both European honeybees and indigenous bees 4. Implement standards for nucleus colony, queen, and bee products both for indigenous bee species and European honeybee 5. Create a favourable environment for the entrepreneur who wishes to engage in beekeeping 6. To establish multiplier farms in beekeeping (Mellifera and native bees) managed by beekeepers, LGUs, and academe	Royal jelly production conducted ➢ Training on product processing and packaging provided to beekeepers ➢ Attendance to beekeeping-related activities like beekeeping conferences, techno-for an exhibit, educational tours, etc. • Implement Best Management Practices in Beekeeping standards as approved by BAFS • Production of beekeeping reading materials, C.D, market linkage, and establishing networking • Multiplier Farm established
C. To promote, develop, and commercialize the raising of indigenous bees increasing diversity and protecting them from extinction	1. To new and old beekeepers in the management and production of indigenous bee species 2. Encourage other farmer associations, private and state universities, and LGUs to launch beekeeping projects	• Beekeepers are trained in the management of indigenous bees. • New institutions involved in beekeeping programs.
D. Biosecurity measures from major and fatal pests and diseases of bees	1. To train quarantine/deputize officers in the detection of pests and diseases 2. To train staff of the Regional Disease Diagnostic Laboratory (RADDL) in D.A Mandaue and Molecular Laboratory in Ubay Stock Farm in Ubay, Bohol in laboratory examination of diseases 3. Create partnerships in monitoring and regulation of pests and diseases between the D.A Bureau of Animal Industry quarantine	• Quarantine/deputized officers trained • The staff of RADDL and Molecular Lab. is trained in the detection of diseases. • Committees involving different stakeholders were created to monitor,

	service, beekeeping assoc., academe, and LGUs involved in beekeeping activities 4. Laboratory research activities were conducted to prevent, protect, and eradicate major pests and diseases.	and regulate pests and diseases. • Laboratory research activities were conducted.
D. To establish a Program Mgt. The team (PMT) and a different committee will strengthen the implementation, monitoring, and evaluation of the Road Map	1. Organization and Activation of Program Management Team (PMT) and committee with its corresponding functions and responsibilities among different beekeeping stakeholders	• Program Mgt. Team and different Committees created and functions delineated

4. Omoyon Bee Farm, Dumaguete, Megros Or.
5. Melvin Guerero, Matlan, Isabel, Leyte
6. Noel Regulacion, Ormoc, Leyte
7. Nestor Evaristo (Aboy) , Don Salvador Benedicto, Negros Occ.
8. Alberto Bartolata, Pangdan, Cebu
9. Nathan Amizola Magsayo, Tubigon, Bohol
10. Narciso Jalop Jusay Jr., Corella, Bohol
11. Josha Lupian, San Guillermo, Palompon, Leyte
12. Balag Jomer, Jolit's Eco Farm, Batuan, Bohol
13. Jhune Baluca Tidalgo, Tagbilaran, Bohol
14. Jonji Jesus, Molo, Iloilo
15. Alfredo Gulmatico, Cabatuan, Iloilo
16. Pamela Henares, Bacolod City
17. Dodjie Marquez, Bacolod City
18. April Fatima Dira Villanueva, San Joaquin, Calbayog, Samar
19. Jezrael Vergara, Pajo St., Lapu-lapu City
20. Glenn Estiva Lagumbay Oslob, Cebu
21. Ruel Benitez Delantar, Maypangdan, Borongan City, Eastern Samar
22. Meredith Flores Arillo, Sitio Buntalan, Poblacion, Leon, Iloilo

APPENDIX 2 - SEASONAL MANAGEMENT OF STINGLESS BEES IN THE SOUTHERN REGION OF CEBU

The Climate of Cebu is a tropical wet and dry climate. There are two seasons in Cebu - the wet season and the dry season, but Cebu has three different climates, based on the distribution of rainfall. The city of Naga belongs to the southern part of Cebu. The months of February to May have the lowest rainfall and coincide with the summer or the dry season resulting in most trees blooming and nectar flowing.

June to September is the monsoon season but several crops bloom during this month and the flushing of leaves of most seasonal trees occurs then. It is also the time when pollen flow can be observed in most colonies. October to January has a minor flow of both pollen and nectar.

Production of pollen and nectar will be dependent on climatic conditions, vegetation within the area and the management efficiency of the beekeeper. The environmental factors that affect production are soil moisture & fertility; temperature; humidity; day length and sunlight.

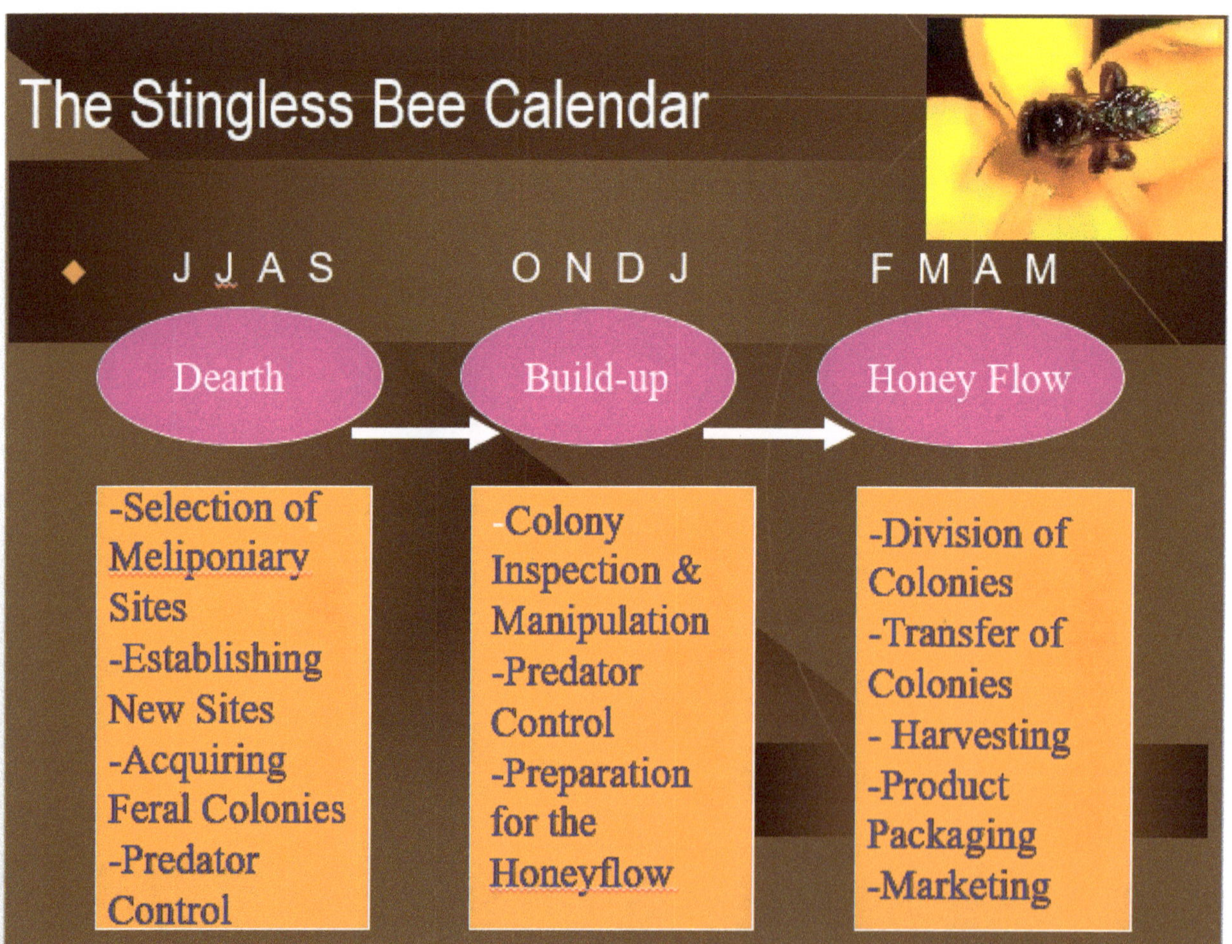

Chart 12 The Visayas Stingless Bee Calendar

Stingless bees adjust their schedules to coincide with the plants' daily rhythm which is also affected by the climate. Based on my experience for the past six years raising stingless bees and 30 years of being a Mellifera beekeeper, my preferred seasonal management is presented in the matrix below:

The exact location of SB enthusiasts in Cebu under the network of BEES Assn.

1- Alberto Bartolata- Pangdan, City of Naga, Cebu (250 colonies)
2- Dionisio Panilagao- Poblacion South, City of Naga, Cebu (10)
3- Matias Tabasa- Poblacion, Bogo City, Cebu (22)
4- First Consolidated Cooperative in Tanon Seaboard, Tuburan, Cebu (20)
5- Clavel Aves, Poblacion Tuburan, Cebu (2)
6- Peter Madjus, Bulacao, Cebu City (30)
7- Arnel Fernandez, Consolacion, Cebu (4)
8- Rolando Jimenez, Campangga, Barili, Cebu (2)
9- Chavez Chavez, Poblacion, Compostela, Cebu (2)
10- Arnel Fernandez, Poblacion, Consolacion, Cebu (4)
11- Jezrael Vergara, Pajo St., Lapu-lapu City (2)

APPENDIX 3 - COEXISTENCE OF SB COLONIES IN ARBOREAL ANT AND TERMITE NESTS IN THE PHILIPPINES

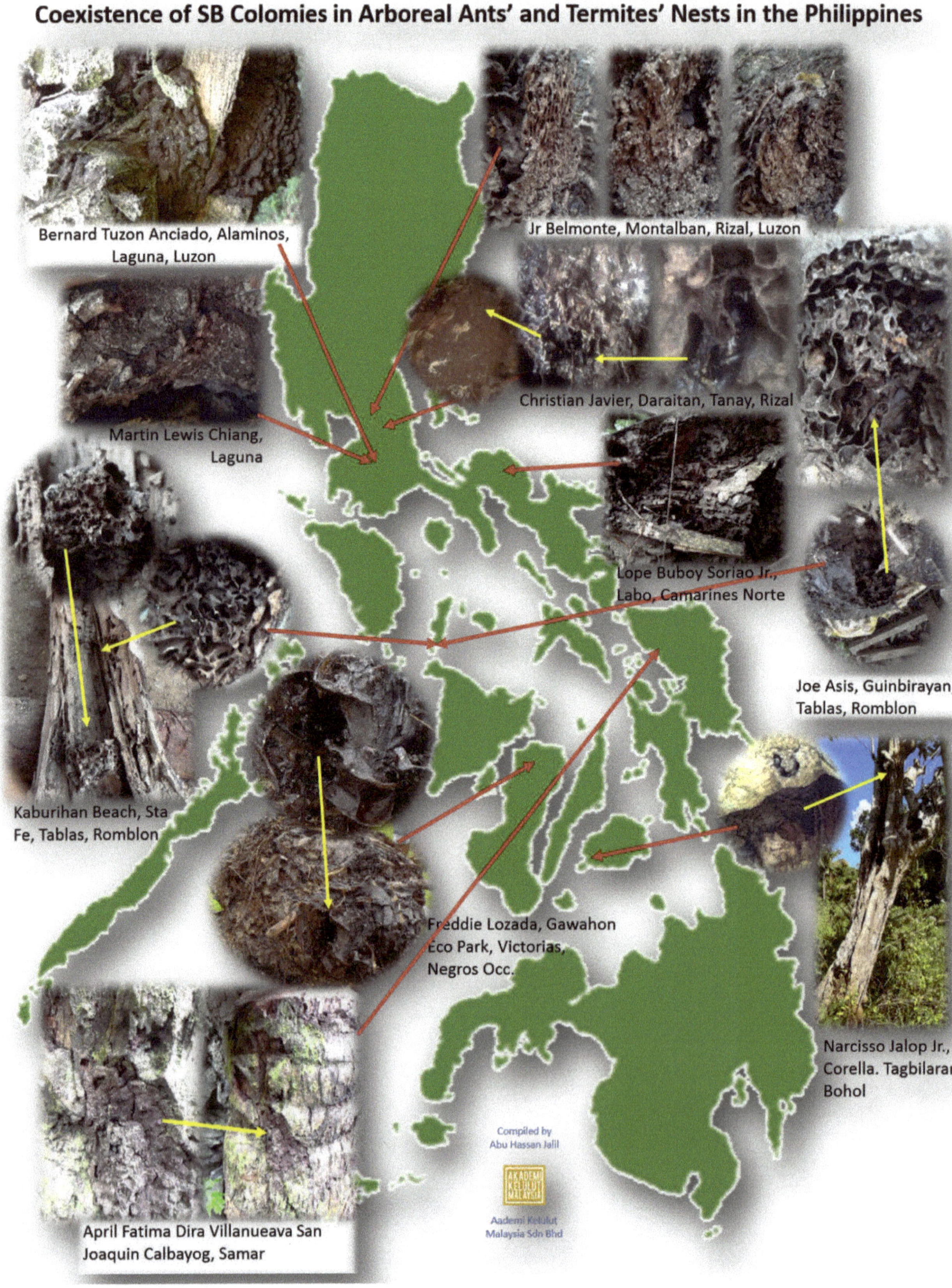

Chart 13 Coexistence of SB Colonies in Arboreal Ants' and Termites' Nests in the Philippines

Reports from Eraldwin Dimailig in Lucban Quezon and Marc Lyndon Genvana in Jordan, Guimaras (without photo records)

Comments of Melvin Monzon "What is special about the coexistence? Of the hundreds of colonies I've extracted from the forest, I'd say that is one in a hundred where I'd see a termite or ant colony beside a stingless bee colony. That is not the norm, but rather, just a coincidence. If ever there is a symbiotic relationship between a stingless bee colony and a termite or ant colony, they should be beside each other always, like the remora to a shark.

I just recalled a co-occurrence of *T.biroi* and termite in a dead but still-standing sugar palm tree. The sugar palm tree is highly susceptible to termites. It's just a coincidence that the sugar palm tree is a host of the fern in which the *T.biroi* nests. While the tree is still alive, it is host to the fern we locally call 'dapo', and many dapo grow on the tree and this nesting site is used by *T.biroi*. But once the sugar palm has fruited, it'll die and a termite colony will then move in. However, it does not appear to be any symbiotic relationship between them.

If we can separate/isolate the stingless bee colony from the termite/ant colonies, and the stingless bee colony not just lives but thrive, that only means there is no relationship, there is no dependency between the two species. It's just a coincidence, a one-in-a-hundred scenario, that those colonies are sitting beside each other.

There is this one time when there was an eagle's nest situated on top of a *T.biroi colony*. That is not the norm, it's just a coincidence. In the scenario where an ant nest is adjacent to a stingless bee colony, the moment the stingless bee colony is damaged and exposed and the damage is not repaired immediately, it'll be assaulted by its ant neighbour. I don't see anything relevant to this coexistence. But if I'm missing something, please do enlighten me."

Difference between Cohabitation and Coexistence

Living in the same tree is **co-inhabit**, but when colonies occupy the same nest shell they coexist. In the Philippines, the encounters were found in the same tree, and the term coexistence was mistakenly applied. Only in the Greater Sunda Islands encounters were some nests in the same nest shell. In subterranean termite nests, the SB don't cohabit but replace the termites as occupants of the host mound. I have not seen SBs replace termites in arboreal nests. When termites get cellulose they rarely leave, and SB can live nearby. Termites will leave only when cellulose is exhausted.

Here all these coexistence studies are on feral or wild colonies.... in domesticated hives, the ecology is different. This is not a study on domesticated hives or commercial bee farms. In domestication or commercialization, the termite is a pest and enemy, pure and simple. There does not necessarily have to be a relationship between termites and SBs when they coexist.

Significance

This message has prompted me to highlight the significance of understanding co-habitation.

1. It indicates that the SB nest shell is impenetrable by termites due to the properties of the batumen of the SB nest. From this fact, we now know that extracts of the batumen itself can be used to protect our wooden furniture.
2. The dead trees infested by termites are bound to fall sooner or later, sooner in the high winds of a typhoon. Those that have SB nests coexisting with these need saving and hence add to the collection of a nearby beekeeper.
3. Mapping or charting its occurrence, tell hive hunters they need not ravage forests in the hunt for wild SB colonies as they are more likely available in termite-infested trees, indirectly this may lead to the conservation of SB colonies in live trees.
4. Every time there is a typhoon, beekeepers can look at the brighter side of calamities and go into the wilderness to gather fallen branches in the hope of encountering termite-infested trunks or branches with SB colonies—optimism at its height.
5. This chart will show the widespread coexistence and beekeepers need not worry about the presence of termites on their farms. Many a time, beekeepers try to eradicate termites with pesticides but end up killing their own SBs as well. This chart will show they coexist peacefully because the termites cannot penetrate the SB nest shell.

Figure 341 A termite is stopped in its tracks by the nest entrance cerumen.

The termite and SB co-habitation chart is for the whole Philippines, so its significance should be appreciated by all islands. Following here are some translations of the chart's significance in the different dialects.

APPENDIX 4 - *TETRAGONULA* NEAR *BIROI* BY D. R. SMITH

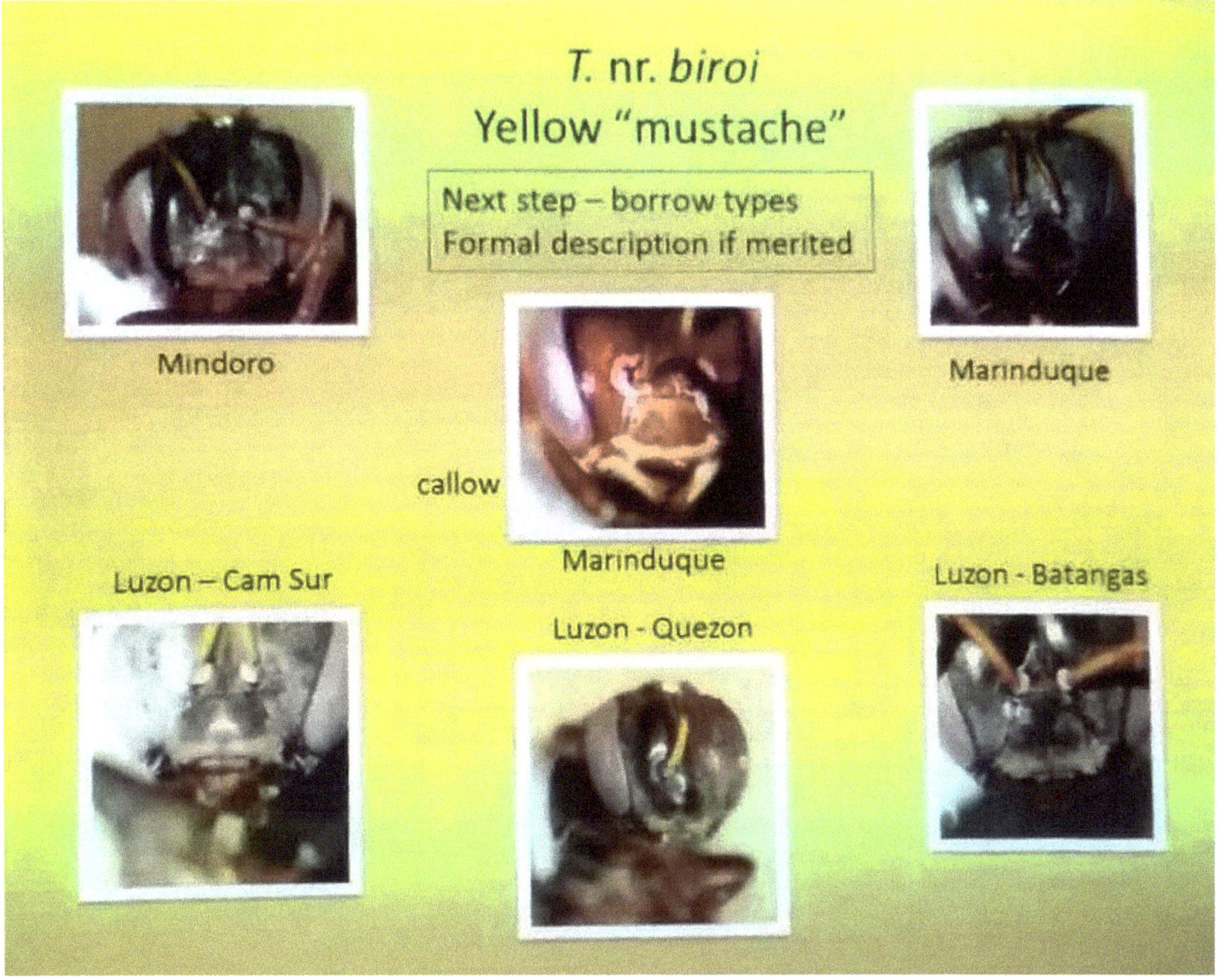

Chart 14 Presentation slide during the conference, courtesy of Prof Deborah R. Smith

In her Keynote address "Genetic Diversity of Stingless Bees" during the AAA 2020 conference, Prof Dr Deborah R. Smith elaborated on the genus *Tetragonula* as an example to illustrate the use of genetic tools by several authors to study diversity. *Tetragonula* species according to her are perhaps the most difficult Asian stingless bees to identify, and many cryptic species remain to be identified.

Image 83 facial profile of Batangas specimen

A collection of 53 Philippine *Tetragonula* colonies was used to demonstrate the use of DNA barcoding as an aid in distinguishing Tetragonula species and to compare differentiation[33] among conspecifics residing on the major Philippine islands. One of them is suspected to be a species endemic to the Philippines and will likely receive a name in the near future. I noticed a lot of variation in the size of the moustache, but no geographic pattern - but I only had a few nests from each island sampled. I don't know if there is really a geographic pattern, but my sample sizes were just too small to pick it up, or if the variation is due to the age of the individual, or even if the variation is simply by family (nest). (pers. comm. D. R. Smith, 2020)[34]

APPENDIX 5 - IDENTIFICATION OF WALLACEA STINGLESS BEE WORKERS

ID 1.0 Illustrated Species Identification

The species in Wallacea has a peculiar range. Although some of the *Tetragonula* overlaps in the Indo-Malaya clade some are from the Australasian clade as well. This calls for a set of keys specific to the Wallacea region.

Figure 342 Palawan Stingless Bees

ID 1.1 Key to workers and comparison charts

Key to *Tetragonula* workers in Wallacea (Starr & Sakagami 1987) (Dollin, Dollin & Sakagami 1997)

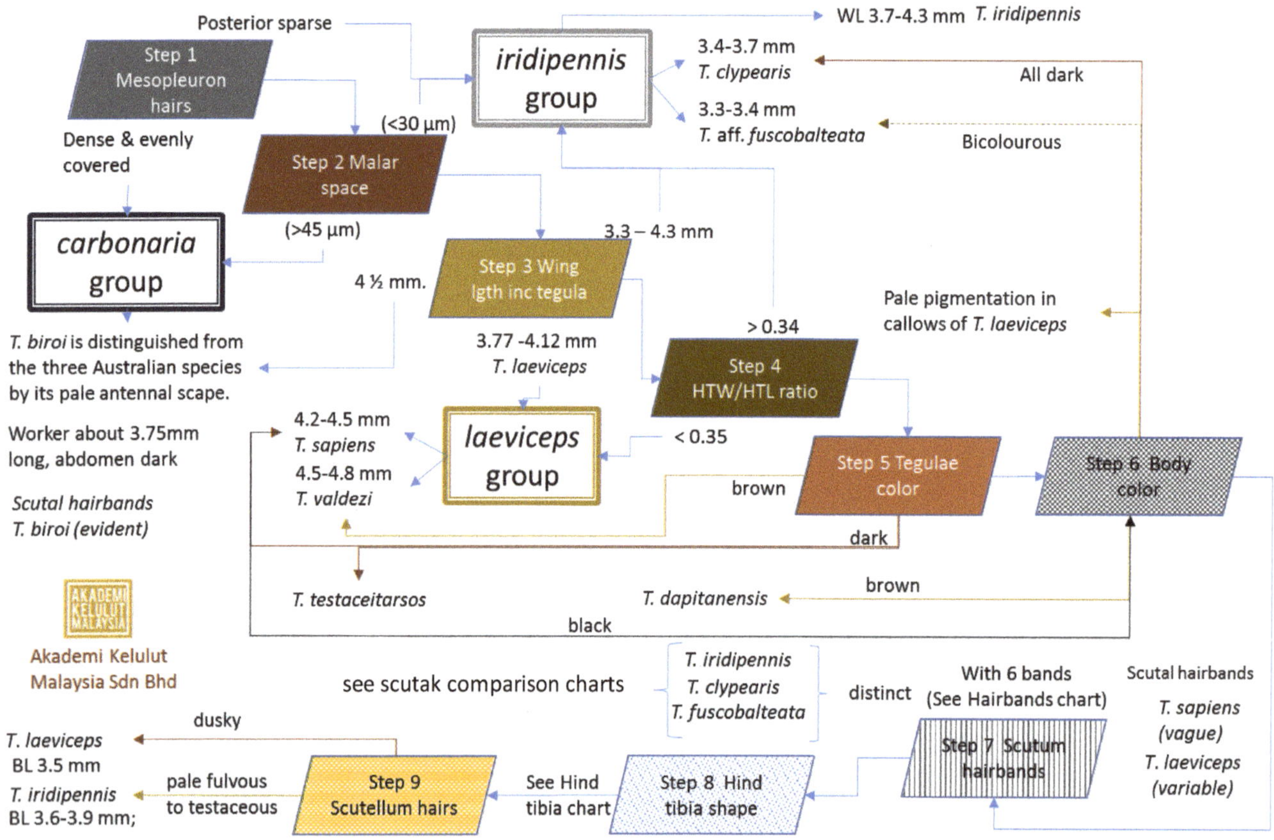

Lepidotrigona nitidiventris* Vs *L. palavanica

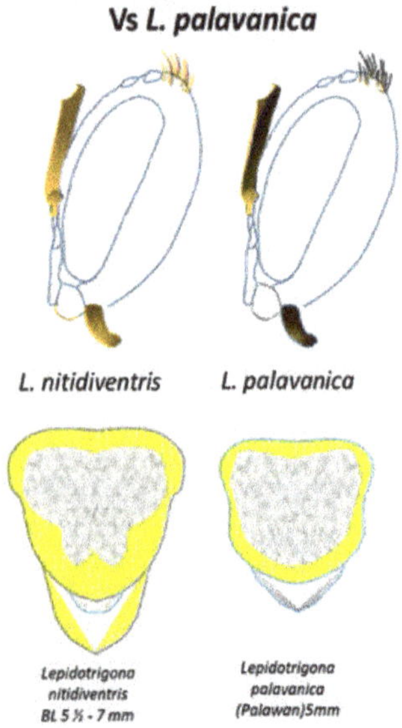

Mesoscutal Hairbands comparison chart for *Tetragonula iridipennis, T. fuscobalteata* & *T. laeviceps* workers (Redrawn from Sakagami, 1978)

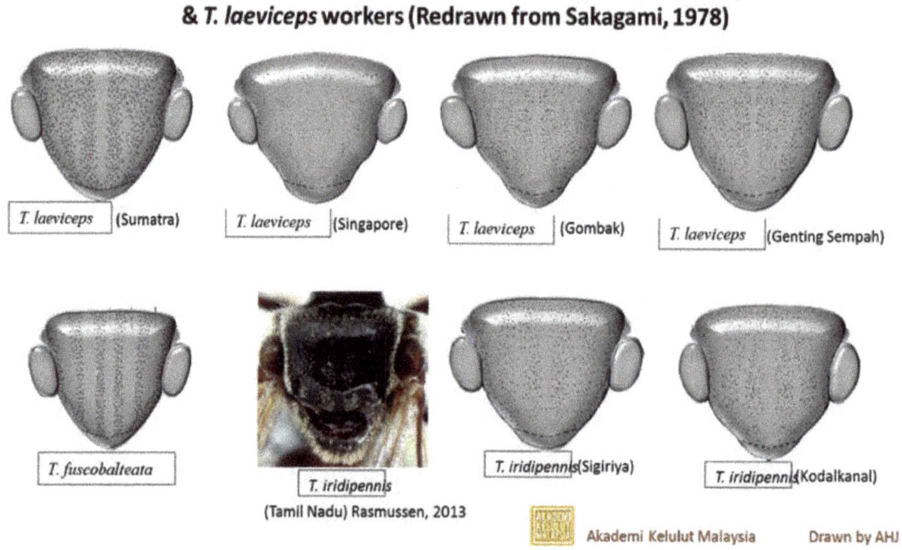

Chart 15 Key to Tetragonula workers ain Wallacea and Scutal hairs chart

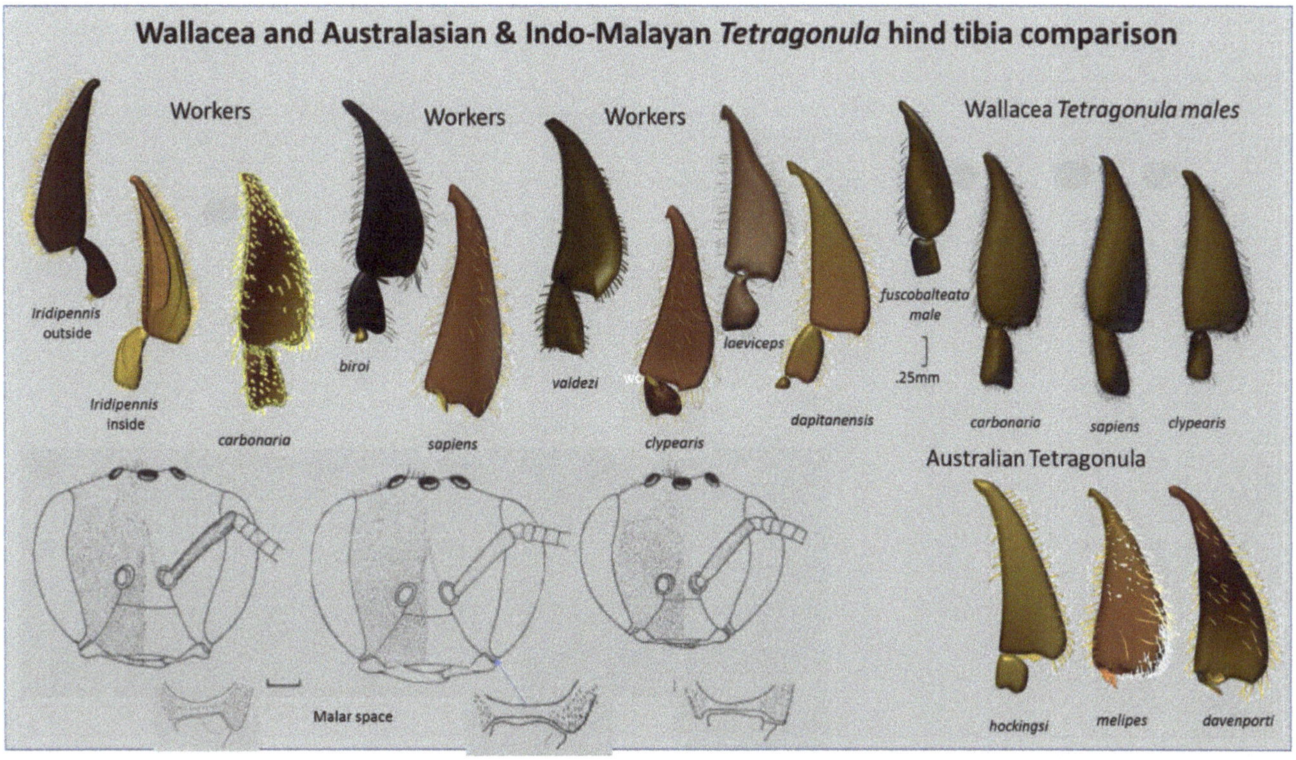

Chart 17 Tetragonula hind tibia comparison chart

Comparison *Tetragonula iridipennis* Smith, 1854 to *T. laeviceps* Smith, 1857

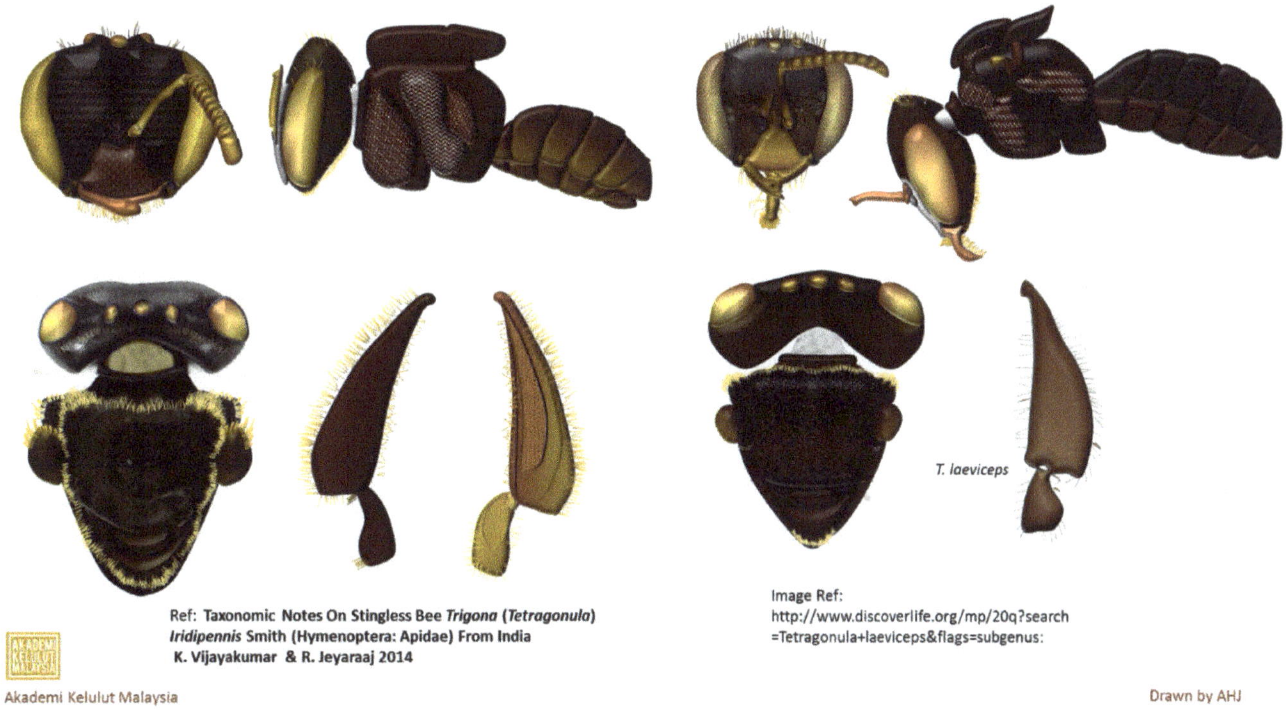

Chart 16 Comparison of T. iridipennis to T. laeviceps

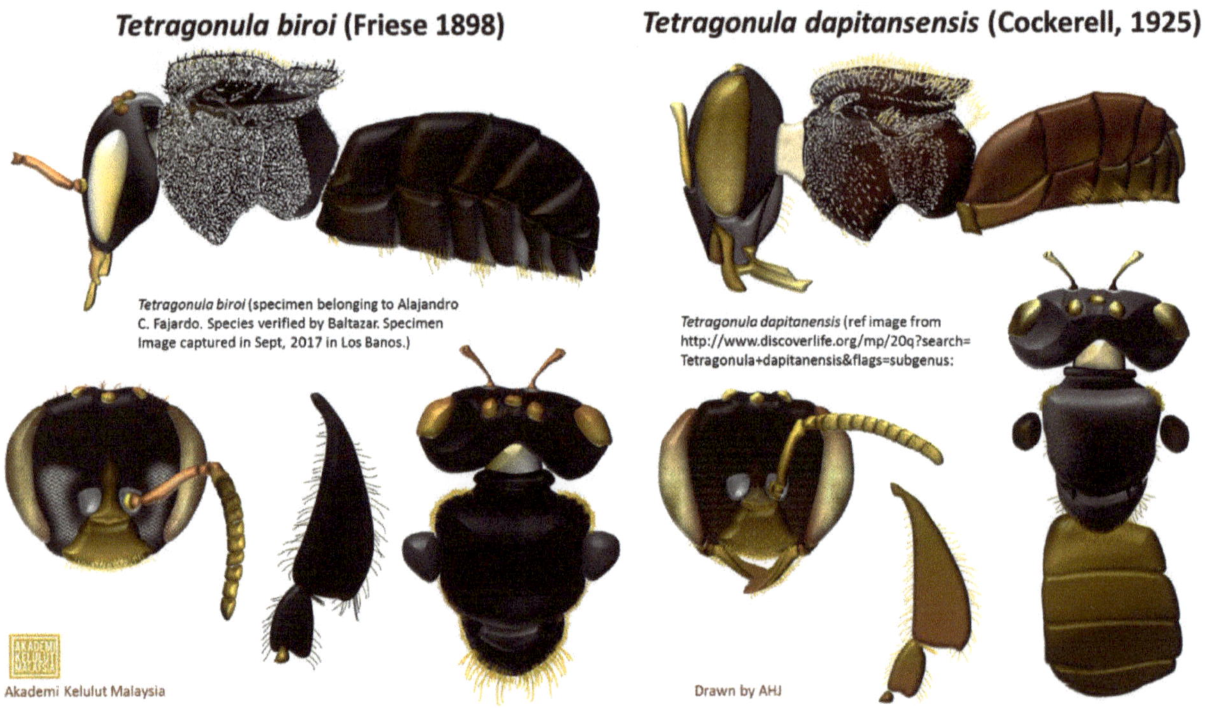

Chart 18 Comparison of T. biroi to T. dapitanensis

Chart 19 Comparison of T. sapiens to T. clypearis

ID 1.3 References

- Ascher, J., & Pickering. (2018, Feb 28). *Discover Life bee species guide and world checklist*. Retrieved from *Wallacetrigona*: http://www.discoverlife.org/mp/20q?search=Wallacetrigona&guide=1&flags=HAS:subgenus
- Ascher, J., & Pickering. (2018, Feb 29). *Discover Life bee species guide and world checklist*. Retrieved from *Tetragonula*: http://www.discoverlife.org/mp/20q?search=Tetragonula&guide=1&flags=HAS:subgenus:
- Baja, S. M., & Mostoles, M. D. (2017). *Colony Characterization And Amplicon Sequencing Of Stingless Bees (Tetragonula biroi)*. In A. H. Jalil, *Handbook of Meliponiculture* vol 2. Akademi Kelulut Malaysia.
- Baltazar, C. R. (1966). *A Catalogue Of Philippine Hymenoptera*. Pacific Insects Monograph 8.
- Chuttong, B., & Burgett, M. (2016). *Biometric Studies Of The Stingless Bee Tetragonula laeviceps Species Complex (Apidae: Meliponini) In Se Asia (Thailand)*. 1st International Meliponine Conference. 2017
- Cockerell, T. D. (1919). *The Social Bees of The Philippines Islands*. The University of Colorado.
- Dollin, A. E., Dollin, L. J., & Sakagami, S. F. (1997). *Australian Stingless Bees of the Genus Trigona (Hymenoptera: Apidae)*. CSIRO Australia Invertebrate Taxonomy.
- Engel, M. S., & Rasmussen, C. (2017). *A new subgenus of Heterotrigona from New Guinea (Hymenoptera: Apidae)*. Journal of Melittology.
- Freise, H. (1898). Die *Trigona-arten* Australiens. Termeszetrajzi Fuzetek 21:, 427-431.
- Inoue, T., Salmah, S., Abbas, I., & Yusuf, E. (1985). 1985. *Foraging Behavior of Individual Workers and Foraging Dynamics of Colonies of Three Sumatran Stingless bees*. Res. Popul. Ecol. 27:, 373-392.
- Jalil, A. (2016). *Meliponine Identifier Pictorial Guide*. Akademi Kelulut Malaysia.
- Michener, C. D. (2007). *The Bees of the World 2nd Ed*. The John Hopkins university press.
- Mostoles, M. (2013). *Cultural Management Practices of stingless bees in Bico*l. CBSUA, Extension Division.
- Rahmanl, A., Das, P. K., Rajkumari, P., Saikia, J., & Sharmah, D. (2015). *Stingless Bees (Hymenoptera: Apidae: Meliponini): Diversity and Distribution in India*. International Journal of Science and Research (IJSR), Volume 4 Issue 1.
- Rasmussen, C. (2008). *Catalog of the Indo-Malayan/Australasian stingless bees (Hymenoptera: Apidae: Meliponini)*. Zootaxa 1935.

- Rasmussen, C., Thomas, J. C., & Engel3, M. S. (2017). *A New Genus of Eastern Hemisphere Stingless Bees (Hymenoptera: Apidae), with a Key to the Supraspecific Groups of Indomalayan and Australasian Meliponini.* American Museum Novitates.
- Roubik, D. (2006). *Stingless bee nesting biology*. Apidologie 37.
- Sakagami, S. F. (1978). *Tetragonula Stingless Bees of Continental Asia and Sri Lanka (Hymenoptera, Apidae).* Jour. Fac. Sci. Hokkaido Univ. Ser. VZ, 2001. 21(2).
- Sakagami, S. F., & Inoue, T. (1985). *Taxonomic Notes on Three Bicolorous Tetragonula Stingless Bees in Southeast Asia*. Kontyii, Tokyo, 53 (1): 174-189.
- Sakagami, S. F., Inoue, T., & Salmah, S. (1990). *Stingless Bees of Central Sumatra.* Sapporo; Hokkaido University Press.
- Starr, C. K., & Sakagami, S. F. (1987). *An Extraordinary Concentration of Stingless Bee in the Philippines, with Notes on Nest Architecture (Hymenoptera: Apidae: Trigona spp).* Insectes Sociaux Vol 34.

APPENDIX 6 - ADVANCED MELIPONICULTURE TECHNIQUES COURSE

AM.1.0 - 1st Day – Extended Night Class[35]

Figure 343 The propolis preparations extended into the night because they had to complete the immersions for the overnight process

Figure 344 The participants were teamed into groups of three and each had a specific solvent to work on

AM.1.1 – Propolis preparation[36]

AM.1.2 – Solvent preparation[37] *& Product Development*[38] **on the 2nd Day**

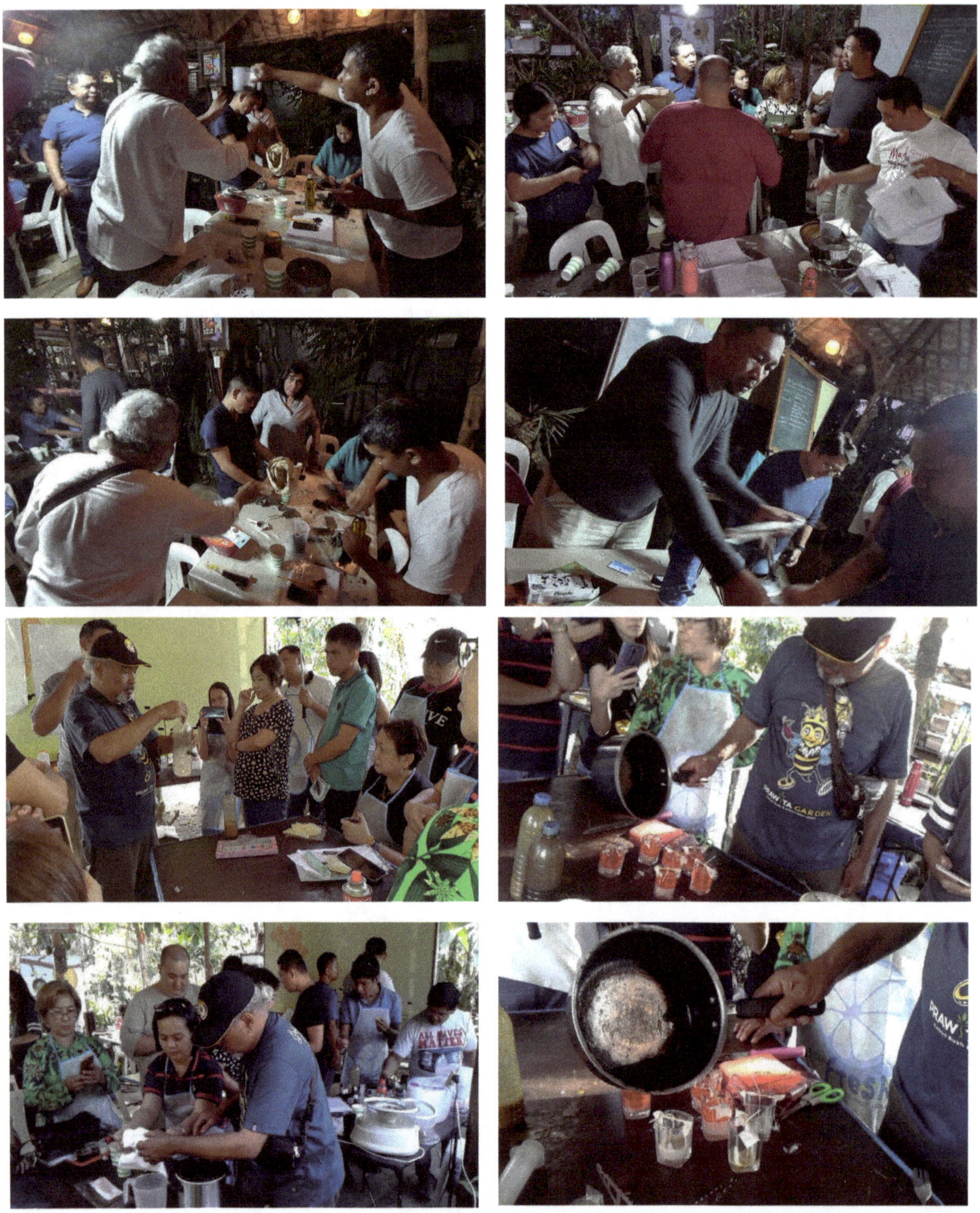

Figure 345 The second day saw the actual processes of the product development

---

[35] There was a power failure earlier and the class had to extended into the night for the Propolis preparation.

[36] See 6.3.1 Types of Propolis.

[37] See 6.3.4 Suggested extraction methods.

[38] Some of the course information - 6.2 Advanced Meliponiculture: - course excerpts.

AM.1.4 Steam Distillation of Propolis[39]

Figure 346 Glass distilling apparatus set up for extraction of propolis hydrosols

[39] Seen here is a fragile glass distillation setup whereas a stainless-steel setup can be used for safer precautions in situations of robust handling among participants.

AM.1.5 Photos from Lee Gaitana, the course host and owner of the premises, Pia's Bee Farm

Figure 347 Books displayed for reading during breaks and the group photo of the course participants

AM.2.1 Advanced Meliponiculture: - excerpts

AM.2.2 Outline The course outline was as follows

Day 1: Upgrading from Beekeeping to meliponiculture

a) Colony Multiplication and splitting the Brood Chamber; Eduction & swarm traps
b) Pests and Predators – Preventative Maintenance
c) Meliponiculture using principles of Beescape
d) Basic Meliponine ID and Wallacea species recognition
e) Meliponary Farm Management
f) Preparation of propolis for overnight extraction

Day 2: Uses of 5 types of propolis in 5 methods of extraction

a) Product derivatives with Ethanol Extracted Propolis (EEP)
b) Product derivatives with Glycol Extracted Propolis (GEP)
c) Product derivatives with Oil Extracted Propolis (OEP)
d) Product derivatives with Turpentine Extracted Propolis (TEP)
e) Product derivatives with Steam Distillation or Aqueous Extracted Propolis (AEP)

(Ref: Krell, R. (1996). Value-Added Products From Beekeeping. Rome: FAO Agricultural Services Bulletin No. 124)

AM.3.0 Day 1 Course

AM.3.1 Types of Propolis

There are 5 types of propolis

a) Batumen is the hard cavity wall linings or the shell of a SB nest. This keeps intruders out.
b) Involucrum is a cerumen sheath enveloping the brood and may consist of many layers to regulate and stabilize the brood temperature and humidity.
c) Cerumen of the nest entrance tube or funnel, with resinous 'ornamental' spikes or flanges.
d) Storage pots. Also built with cerumen, they hold honey, pollen, rarely wax or resin, or others…
e) Pillars are cerumen that holds all components in place and also supports the components as struts

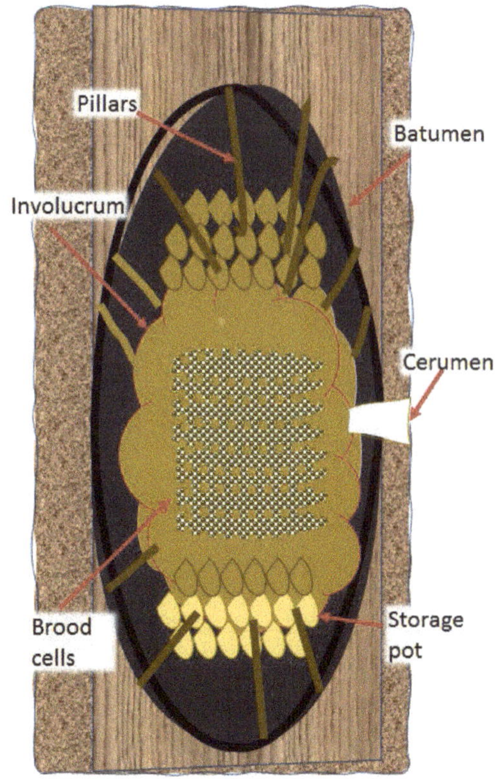

Chart 20 Types of Propolis

AM.3.2 Tree Hollows

Figure 348 Some natural cavities and some made by termites

A tree hollow is a cavity in a living tree. Usually, when trees grow large enough the core or bole of the tree may naturally have a cavity and some subterranean insects that exploit this from the buttress up, internally. Some trees lose lower limbs or branches that break off or have rotted through weathering, which creates a cavity in the trunk. This is where the stingless bees may gradually chew their way in.

On other occasions, termites have made some cavities and the bees take advantage of this but make a barrier with batumen and then expanding their nest in those cavities.

Generally, stingless bees would prefer a living tree with the cambium layer intact as it provides insulation and protection. However, a cavity in a rotting termite-infested tree may frequently lure them. Tree ferns have fibrous trunks which are hollow and house stingless bee colonies in a similar manner. Most tree ferns are members of the "core tree ferns", belonging to the families Dicksoniaceae, Metaxyaceae, and Cibotiaceae in the order Cyatheales. These have been found in the Eastern regions of the Philippines down to Samar and Leyte-Sorsogon.

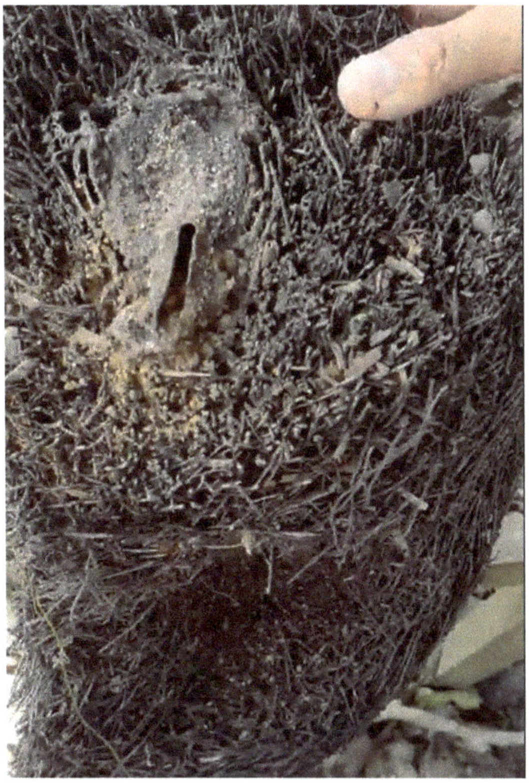

Figure 349 Core fern tree houses a stingless bee nest in Tacloban, Leyte

In certain cases, a cluster of epiphytes may also provide a cavity large enough to host a colony of stingless bees. This is mostly found in the Bicol region and acquiring them may require some skill. Although sometimes during high winds or typhoons, these colonies may "drop from the sky", so to speak.

AM.3.3 Limestone rock cavities.

Limestone is a sedimentary rock **composed** largely of the mineral calcite (calcium carbonate: $CaCO^3$). Limestone rock cliffs or outcrops after many years of exposure to the weak carbonic acid in rainwater, reacting with the chemicals in the rock, dissolved and eroded the limestone. Large hollow solution cavities were formed in the limestone in this way. These cavities are then exploited by stingless bees. These occurrences may manifest in the Palawan and Mindanao limestone cliffs.

The nest architecture and basic structure are similar to the nest structure in the tree hollows or cavities. All stingless bee nests will have the five main types of propolis. It may differ in some species or depending on the cavity size or the material and texture of containment walls. In all instances, it is advisable to harvest or collect the different types separately to process them separately for different derivative products.

AM.3.4 Suggested extraction methods

Table 3 Tabulation of Propolis type against suggested solvents

No.	Propolis Type
A	Batumen
B	Involucrum
C	Nest structure excluding Batumen & entrance Cerumen[40]
D	Storage pots & pillars
E	Entrance Cerumen only

No.	Suggested Solvents		Code
1	Ethanol	Ethyl alcohol or Isopropanol	EEP
2	Glycol	Propylene Glycol or Vegetable Glycol (Glycerol)	GEP
3	Oils	VCO; Olive; Soya; Carotino (Red Palm Olien); Rice Bran; Sunflower seed	OEP
4	Aqueous	Steam Distillation	AEP
5	Turpentine[41]	Alternative: Linseed or Flax	TEP

AM.3.5 Stingless bee sizes

No.	Genus	Species	Size
i	Heterotrigona	i. erythrogastra	xL
2	Lepidotrigona	i. nitidiventris,	L
		ii. palavanica	L
3	Tetragonula	i. biroi	M
		ii. clypearis	xS
		iii. dapitanensis	L
		iv. Fuscobalteata	xS
		v. laeviceps	M
		vi. sapiens	S
		vii. testaceitarsis	L
		viii valdezi	L

Chart 21 Relative Stingless bee sizes

The KEY to the identification of the genera is not too difficult for bees in the Philippines and is quite straightforward; however, the key for species of *Tetragonula* is rather complicated.

"The main difficulty in sorting *Tetragonul*a is the virtual absence of reliable structural characters in workers. Consequently, the classification must depend on the size, proportion colouration and pilosity, which makes the identification of particular specimens difficult or often impossible (Sakagami, 1978)."

As the main character is body size, I classified them into S, M, and L sizes as the initial step. This factor can be obvious if the full range is laid side by side. Since getting the full range is not as easily done as one would expect, I have drawn up an algorithm based on keys from Starr & Sakagami (1987);

[40] Cerumen is the same material used for entrance tubes, involucrum sheaths, brood cells and storage pots. All interchangeable- while pliable and not brittle. The brood cells, those with eggs or larvae, supply a lot of cerumen when the mature larva spins its cocoon, followed by removal of all cerumen from cells by the workers. (DWR)

[41] Turpentine, Linseed oil or Flaxseed oils are suitable for use with batumen or nonpliable and crusty type nest entrance tube to produce wood preservatives, termite repellents and wood finishes and polishes.

Dollin, Dollin & Sakagami (1997); Sakagami (1978); Rasmussen and Michener (2010), and Schwarz (1939).

AM.3.6 Foraging needs

a) Among the SB needs, the colony will require resin, gum or latex to construct its nests.
b) Wax synthesized from its own body is mixed with the vegetative exudates in different proportions to result from the different types of propolis (all sticky vegetable substances) and brood cells.
c) Other needs besides nectar and pollen, in small amounts, are honeydew (processed phloem sap exuded by sap-feeding insects, Hemiptera), salts, minerals, and water and these are obtained sometimes from human or animal sweat and wastewater.
d) They have also been seen foraging seeds or flour, lichen, mould spores and mucilage.
e) Proteins sometimes are foraged from sources like tears, plasma or albumin.

All these foraging needs vary in supply depending on seasons and provisions in the Bee Garden prepared by the Stingless Beekeeper.

In the wet season, one may encounter abundance in volume but not necessarily in concentrated resources. The moisture content will often be higher in wet ambient conditions.

Honeydew and some other relatively dilute sugary liquids from three major sources occur.

a) Dew settles in the narrow flower corolla and dilutes the nectar
b) Extrafloral nectar may have a higher water content
c) Hemiptera may also have increased phloem foraging, thus excreting more honeydew fluids

All this will be the opposite in the dry season. This directly influences the amount of honey production in a Meliponary.

Many other factors that may influence honey production are the flowering season of the type of plants encountered, the cycle of vegetables in farming activities and also the fruiting season of orchards or plantations.

AM.3.7 Yearly Flowering/Fruiting & Weather cycles

When looking at flowering plants on a bee farm one will notice that a majority of the flowers have cycles. There are tropical evergreen plants that may have a monthly cycle and shrubs mostly have a quarterly cycle, similar to annuals found in the temperate climate. Then there is the tropical biannual fruiting cycle and on top of that, there is the regular annual weather cycle of wet and dry seasons.

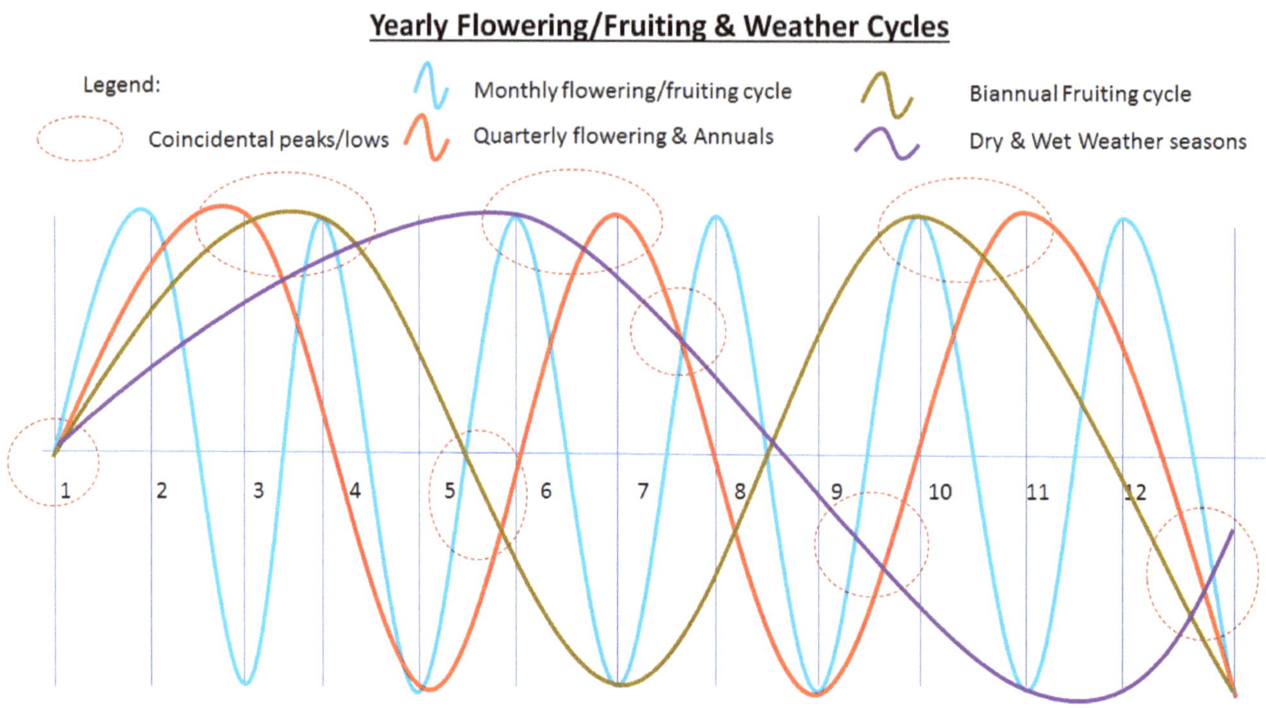

Chart 23 Yearly Flowering/Fruiting & weather cycles)

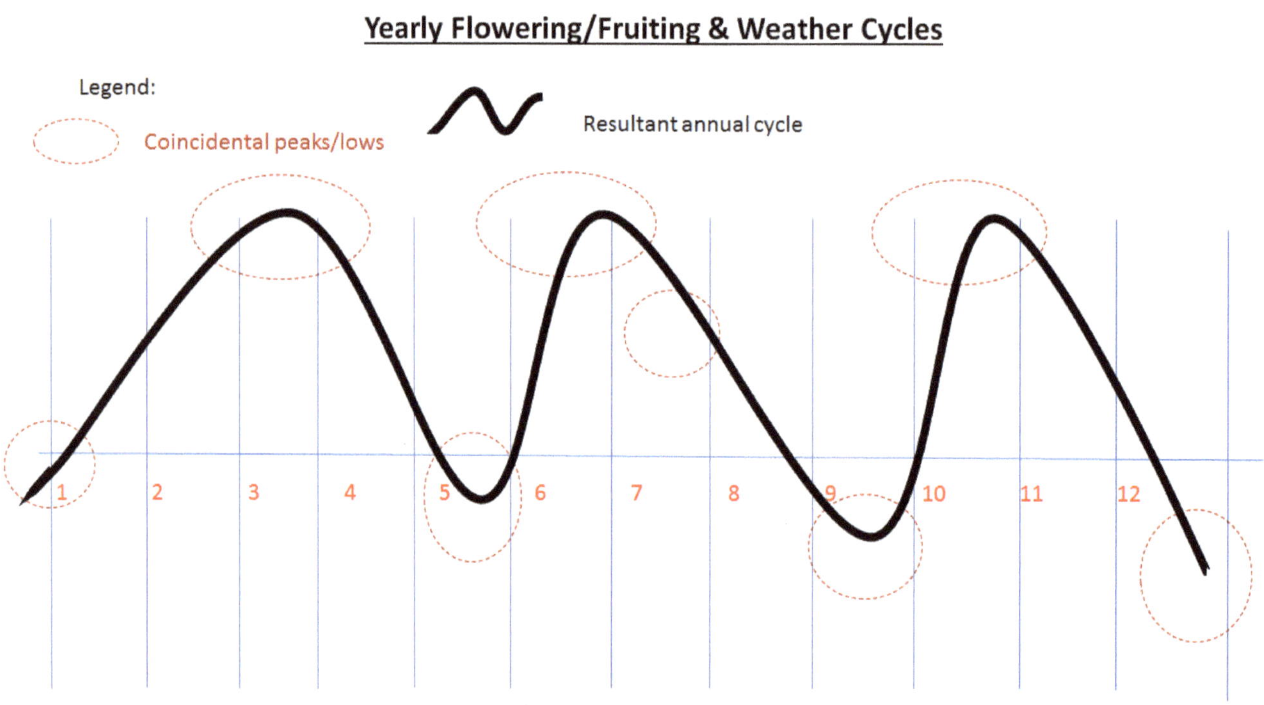

Chart 22 Resultant annual cycle,

Graph[42] Yearly Flowering/Fruiting & weather cycles). A hypothetical situation where all cycles start on the 1st of January

[42] Appended graphs are the cycles that may occur in a plot with diversified planting of a Meliponary or bee farm (see Graph 1 Yearly Flowering/Fruiting & weather cycles) & Graph 2 Resultant annual cycle.

AM.3.8 Contamination Control[43]

Besides locating and gathering their plant resources, the stingless bees forage materials with terpene-based aromatics like paint or stationery gums. It has been observed that they strip off flytrap or lizard trap paper with the glue or gum that the paper is smeared with. They have been known to collect paint particles not only when wet but even old flaking paint and in doing so they tend to pick up rust particles as well.

AM.3.8.1 Paint Contamination

see Mikee Balais

AM.3.9 Replacing a box hive

Replacing a weathered hive box is a major chore not only for the beekeeper but also for the bees. The damage recovery extends longer if not doubled, doing such maintenance activities requires a transit area or sickbay in a Meliponary. It simply will not be achievable if not segregated from the rest of the hives on the bee farm. Such activities of replacing or renewing broken-down boxes need to be minimized, and the way to do that is to ensure that box durability is maximised.

see November

AM.3.10 Hive Box durability

To improve durability in lumber for Hive boxes first and foremost, plywood should not be used for two reasons. One is that the plywood glue will tend to interfere with bee communication because of the formaldehyde contents. And two, plywood does not last more than two yr. in the field. One needs to utilize thick lumber from 1.5 to 1.75 inches to provide the best insulation and minimise warp in the wood. The joints can be rebate jointed with rustproof screws for affixing, instead of nails that will eventually rust. Avoid using ordinary nails that rust, where possible.

[43] Appended also is a gallery of photos of cerumen contaminated with paint flakes and particles (see Paint Contamination overleaf).

Figure 350 Old lumber and 1,75" thick rebate jointed boxes.

Another reason for using thick lumber is to make it heavier so that hives will not be blown off their platform in high winds. Appended herewith, are some examples of the usage of seasoned lumber with the appropriate thickness.

AM.3.11 Securing hive boxes.

To be certain the hives are not swept away by typhoons, it may be necessary to anchor them with their stands collectively into the ground. This way they need not be gathered and strapped down at every impending Typhoon.

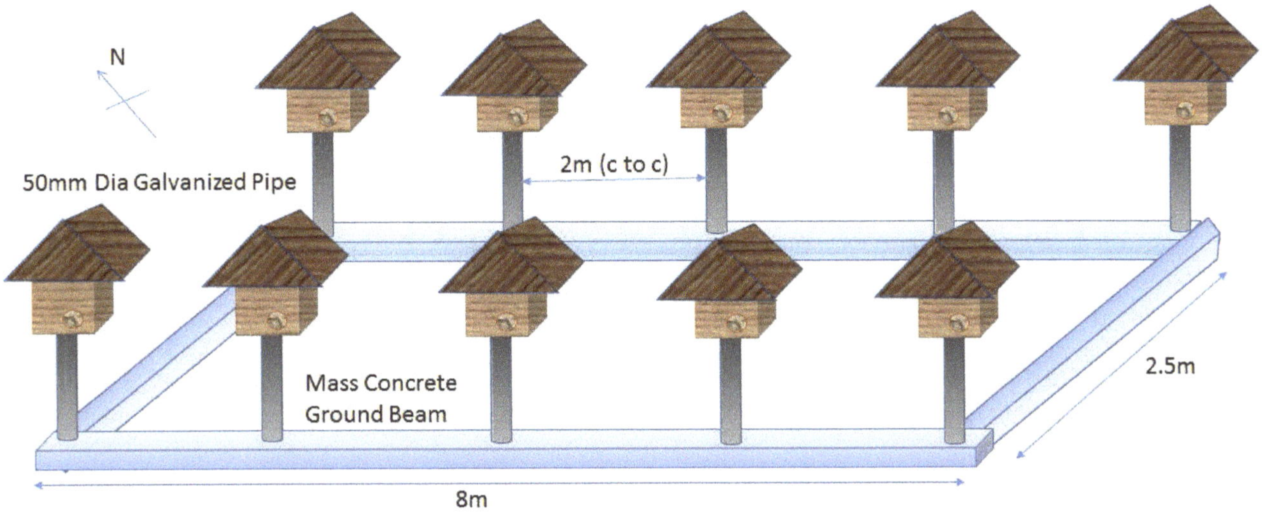

Figure 351 Example; 10 colonies anchored to a rectangular ground beam set about 30 cm in the ground

AM.3.12 Colour assist for returning foragers

To reduce drifting and stop returning foragers from getting into the wrong beehives, the boxes may have various colours and sizes or shapes. Also, importantly, glossy paint should not be used.

AM.3.12.1 Coloured assortment of hive boxes

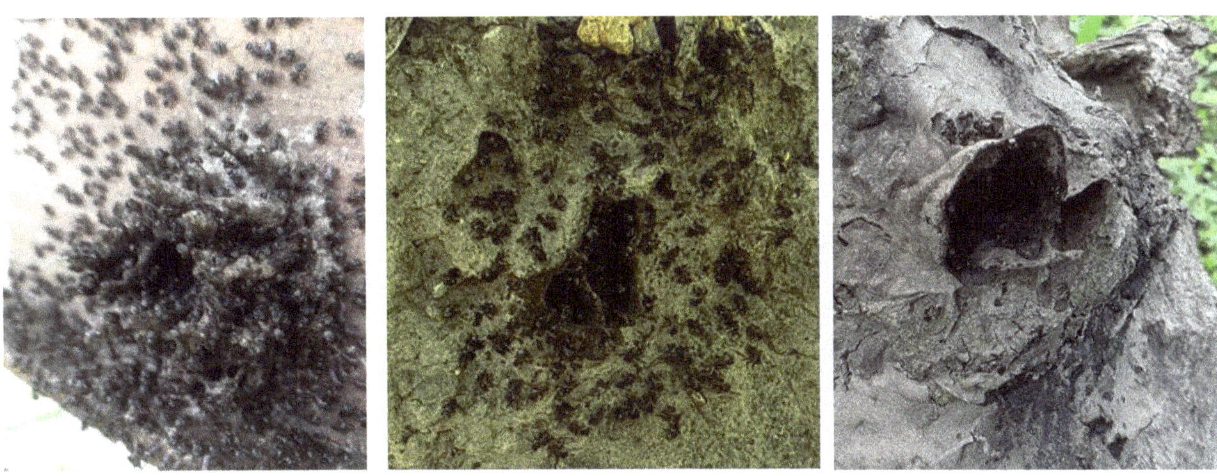

Figure 352 Left: Strong; Middle: Moderate / Regular; Right: Weak – this phenomenon is applicable for species that regularly show congregations outside the entrance

AM.4.0 Colony Health and Strength

Strong healthy colony in Cebu with flared entrance and high FIFO

Normal regular colony in Silay City with full guards on alert.

Weak due to contamination, fungus or parasitic pest

Figure 353 Depiction of colony health by comparison of nest entrance tube or funnel

AM.4.1 Methods of recognizing colony health

a) Entrance tube/funnel aperture
   i) Flaring indicates a strong colony
   ii) Regular constant aperture throughout the tube shows normal good health
   iii) Constricted aperture means the colony is weak, maybe due to some contaminant, fungus or presence of some parasite or pest inside
b) FIFO or fly in fly out activity — the traffic of bees flying in and out through the entrance is a reliable sign of colony population and strength
c) Texture and sheen of the ornate entrance — ornate irregular entrances do not have a single tube or funnel. They may appear as spikes or an ornate matrix. One can tell from the texture and sheen, and overall complexity. A healthy, strong colony will have a shiny moist appearance whereas a weak one will appear dull, lacklustre and dry

d) For certain species, a congregation of bees just hanging around outside the hive near the entrance is a sure way of telling the strength; a weak colony may not have bees there at all.

AM.4.2 Fungus Infestation[44]

It is generally quite difficult to be certain if a colony is infested by just looking at the outside of the hive, however, the lacklustre appearance or aperture constriction, or little activity will prompt one to inspect the interior for moulds or fungi. More likely using containers that are not "breathable" material (e.g. cement, glass, and plastics including PE, PP, PU, PVC etc.) will tend to condense moisture within the hives and this, coupled with mould spores (inadvertently picked up by the bees externally), would ultimately encourage mycelium growth.

[44] Generally, anything that can contribute to mold growth will also attract cockroaches. Cockroaches will eat almost anything, but they love mostly sweet stuff. Moisture control becomes critical when it leads to fungus infestation and subsequently creating an avenue for roach pests to flourish. See appended photos overleaf.

Figure 355 Examples of fungus infested colonies found in the aftermath of Typhoon

"Cucaracha" Family: Blattidae

Fig 7 An added pest and giving the final blow to a deteriorating fungus infested colony are roaches.

Balta notulata "Smoky Brown Cockroach" in Casiguran, Sorsogon

genus Dorylaea? "Harlequin cockroach" Nymphal stage in Santa Fe, Romblon

Neostylopyga rhombifolia_? "Brown Banded" cockroach In Bubog, San Jose, Occ Mindoro

References:

https://en.wikipedia.org/wiki/Blattidae
https://www.google.com/search? =blattidae species
https://www.slideshare.net/Joscella/family-blattidae-2013 (Joscel Enardecido 2013

Figure 354 Examples of fungus infested colonies found in the aftermath of Typhoon

AM.5.0 Colony multiplication

It is a known fact that every colony must have at least one queen. In colony multiplication or husbandry, one cannot duplicate the queen. However, often referred to as colony splitting and with indicators like the presence of sufficient queen cells, many a time real success seems questionable. It is presumed that any of the queen cells will hatch a gyne that will be fertilized to become an ovipositing queen, but this may not be the case.

Leaving this result to chance requires some qualifications. One needs to examine the conditions required so that one-half of a colony without a queen would produce a healthy, viable queen. Having ascertained the colony is healthy and strong, one should now have the criteria to decide on the timing to implement "splitting". The health and strength of a colony, not just the number of queen cells available, are of essential importance in decisions determining the prerequisite to multiply a colony.

Assuming a strong colony population is 10,000, an average colony of maturity would have close to 8,000. If one is splitting an average colony in half, one would be left with two colonies of 4,000 each. This is less than half of a normally healthy colony. The question should now be whether that 4,000 population is sufficient (regardless of the number of potential queen cells) to rebuild the colony back to health. Every split or any operation of domestication requires a "damage recovery period". This is estimated to be at least one generation of workers, which in some *Tetragonula* in Central Sumatra has a minimum of 56 days (Sakagami, Inoue & Salmah 1990). In this case, it may be safe to regard 2 to 2.5 months as the damage recovery period.

Another factor to consider is the condition the newly split or transferred colony is left in while recuperating. It may be detrimental to place the new hives in the masses of older hives on a Meliponary or bee farm. A good bee farm should have an allocated space away from the main farm as an extension or transit area, which can also serve as a "sick bay". This area will be surrounded by sufficient vegetation and flowering groundcovers, flowering shrubs or annuals and trellises with creeping vines and structural shrubs, with a clear water body and resin and mineral sources.

The other major factor is the number of colony multiplications done concurrently at a given point in time. There have been anxious beekeepers carrying out physical duplication of colonies in numbers of 30 in a day. Ambitiously risking the success rate, one can only imagine the result. Many beekeepers assume that if a colony takes within a couple of weeks they have succeeded. Up to three worker generations may be required to reaffirm success. The first generation barely recovers from a physical split because it has to repair the nest structure and rebuild. The 2nd generation concentrates more on adapting to new conditions and environments. This point in time is when new queen cells hatch and the viability of the gyne comes into question. If all ends well, the 3rd generation reproduces, or rebuilds and supports the growing population. It is only the 4th generation that will see the real likelihood of successful multiplication, if only by having one of its drones mate with a queen at a new nest site.

AM.6.0 Honey Dehydrator Schematics[45]

As has been ascertained that stingless bee honey has moisture levels on the high side i.e. above 22% and because this causes fermentation that not only sours the taste. It also allows bacterial growth and provides for too many gases causing bubbling and overflowing foam during packaging and agitation. This led to the need for a dehydration procedure to reduce SB honey moisture content.

.

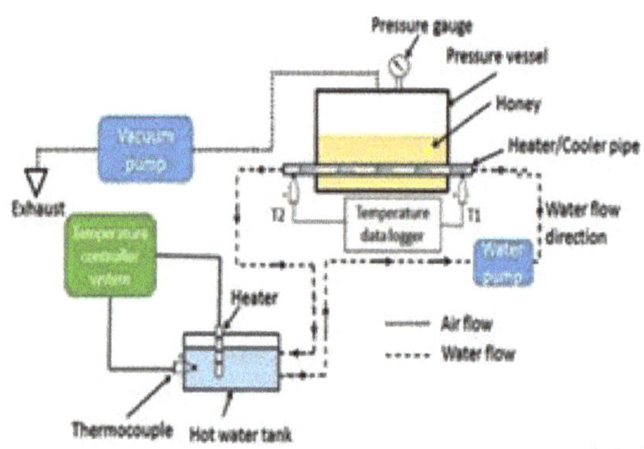

Diagram of a dehydration machine in Malaysia

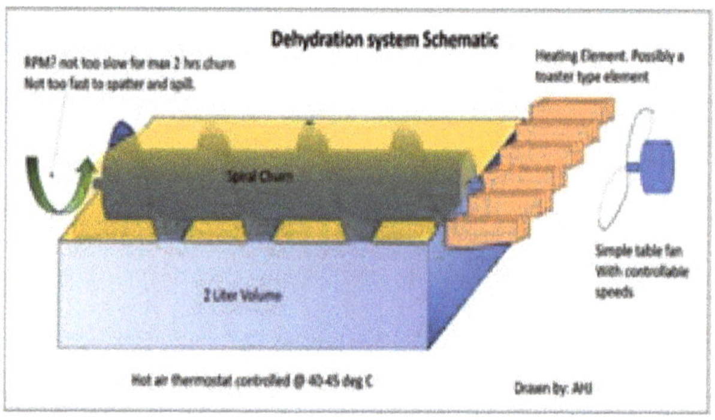

Schematic diagram of a dehydrator inspired by beekeepers in in Thailand on *T. testaceitarsis* and *Lepidotrigona termintata.*

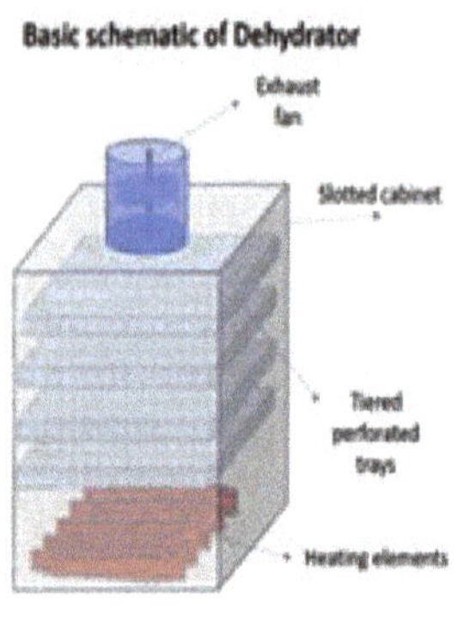

Dehydrator schematics inspired from a contraption by the Fajardo family in Los Baños, Philippines for *T. bitoi* honey

Schematic diagram of a dehydrator inspired from contraptions used in India for *T. iridipennis* honey

Figure 356 Comparing Honey Dehydrators

[45] Appended below are some comparisons of Dehydrator schematics inspired by different countries (see overleaf Comparing Honey Dehydrators).

APPENDIX 7 - TYPES OF SB SWARMS

by Abu Hassan Jalil and Nick Powell

Figure 357 Nick Powell & AHJ

A swarm is defined as a large dense group of flying insects.

7.4.1 Hovering swarm: Expand the defence perimeter and be vigilant to intruders. This is characteristic of a hive with no real intrusion as yet or the presence of a hatchling gyne (virgin queen). It can be from an overpopulated colony venting tidal gas exchange or detecting dummy intruders to real intruders. These hovering swarms may develop into other swarm characteristics as the dangers manifest themselves or the intrusions materialize.

7.4.2 Alert Swarm: This can be when a gyne is due to a hatch sensed from colony pheromones and hive activity. The alert swarm is related to the mating season as opposed to the Alarm swarm which is generally in battling mode.

7.4.3 Alarm Swarm: This happens when the colony is under threat from external impending intruders or danger. These can be sparse initially but grows bigger into attack mode as the imminent danger becomes real.

Can be seen as a tight ball swarm 1-4 meters square with bees' pairs falling and flying back up to join the swarm. This happens when bees grab a friendly bee and are mistaken. This is a key difference between this type of swarm and the attacking swarm. Defensive or alarm swarms can last 1-2 weeks before they feel secure. This swarm is gentler and its pattern while being a vortex is generally not as fast-moving as the attacking swarm

7.4.4 Attack swarm: A follow-up from the Alarm Swarm where the bees come out in full force to attack the current intrusion. This defensive strategy gains its strength in numbers and turns into an offensive strategy. The attack becomes relentless when the brood or its nest produce is ravaged, e.g. bears, badgers, squirrels or primates rummaging for the honey. Other bees or insects rob the propolis.

The swarm contains bees moving in an erratic figure of 8 flight vortex pattern with bees falling out of the swarm in pairs. Bees carry sticky substances on their legs and cling locked to enemy bees locking on with their mandibles to each other. Each pair of bees die as a result.

7.4.5 Invading swarm: Take over manoeuvre by a more aggressive species over a weaker colony. Generally, occurs when a new colony is brought to a farm or due to a lack of or insufficient foraging sources. Sometimes even when the foraging sources are available but not particularly favoured by the newly imported species, the indigenous species may invade to undermine the competition.

This starts usually a few days prior with the arrival of 10-20 scout bees which can be observed hovering in front of the hive entrance. These scout bees test the strength of the colony before the large-scale invading swarm appears in great numbers. The swarm can be noted landing on the hive Turning the face of the hive black with bees. Invaders that land in the hive group together and fan the distressed pheromones into the air to guild the rest of the bees to the location of the invasion. It is a numbers game and the invaders rarely make a gamble on taking a hive.

7.4.6 Mating swarm: Generally, occurs during the dry mating season, from studies in Australia (Halcroft 2012) swarm participants can come from as many as 7 different colonies. Different species in different regions of varied climates or seasons may have different peculiarities.

Drones can be observed in numbers ranging from 20-1000. Hovering and landing on hive roofs and branches nearby to the location of the hive that contains the virgin queen. These can be observed right throughout the year and are more prevalent when ants are sending out their nymph queens. Drone bees hold their antenna in a curved fashion which can be likened to bunny ears. Workers will push drones away from the colony entrance and are on high alert during this time. Hive activity is heightened, and work is done in double time with bees moving very fast in the days before the queen is mated.

7.4.7 Warming swarm: occurs after a spate of cold weather or a period of no sunlight exposure. the swarm usually hover over pockets of sunlight that may be near the hive. These swarms can be as large as 3 meters in width. This was found to occur in Australia after the winter season.

7.4.8 Colonizing swarm: This happens when a colony has reached its capacity. Bees leave the colony in great numbers and cluster on trees, the ground and vertical surfaces in the area they wish to live. Once a location is decided by the swarm workers a pheromone is splattered on the area and continually worked over until the bees in the area that are in flight form together.

All bees will face one direction once the location and the swarm can be seen clustering at the point of entry. This is the only swarm that the bees will all face in one direction in flight as opposed to vortex or figure of 8 flight patterns. Once the location is decided some of the bees will return to the mother hive to take some wax and propolis on their legs. This aids in the fast setup of a new home.

7.4.9 Drifting Swarm: this occurs on a farm when boxes are placed too close and multiple swarms occur, the returning bees may "drift" into the wrong hives in the chaos. This was observed in *T. laeviceps* and *T fuscobalteta* colonies in the west end (Ujong Kulon) of Java Meliponary.

7.4.10 Exploratory swarm: Occasionally seen covering large areas up to 10 meters in a span with bees searching vertically and horizontally usually upon a wall or hedge. Bees landing on and then to rest. Attracted to darker spots of colour variation. Often scent-marking these areas. This can be the initial exploratory move before the colonizing swarm.

7.4.11 Intermingled swarm: This happens when two swarms of different species swarm in close proximity at the same time.

This swarm differs from a fighting or attacking swarm in that the bees are not in an aggressive flight pattern or locking together. The swarm has a soft flight pattern and shows much interest in scent marking. The swarm carries workers and queens of both species and this can change from day to day leaving one more predominant in numbers than the other. Queens can be seen as larger and are a few days old before they join this swarm. Because of the multiple pheromones present in this type of swarm, they are unlike to move into a trap hive and can stay in one area for weeks. Many clusters of bees containing 10-50 bees can be seen working resinous substances along tree branches and vertical walls, posts etc. The size of this swarm can be up to 6 meters in span.

List Of Figures in Part 6

APPENDIX 8 - Directory Of Stingless Bee Filipino Vernacular Names

In our travels, we have compiled some vernacular names for SB. Input from local beekeepers.

1. Kabuyay – clustered brood in Romblon/Sibuyan Island- Sibuyanon and Romblomanon
2. Kadulet - among locals of Macarascas in Palawan.
3. Kalolot – Rinconada, Bikol.
4. Kiwət - Waray of Northern Samar.
5. Kiwot - the Visayan term for stingless bees among beeks[46].
6. Kiyut - In Davao, the term refers to both clustered and layered brood.
7. Kiyut - probably the cute Kiwot of Oriental Mindoro.
8. Kiyut – or kiwot for Tablas Island clustered brood (In Tablas, kabulay means propolis).
9. Lap-og - For layered brood in Sibuyan Island.
10. Lib-og – layered brood in Tablas Island (Manakaran dialect).
11. Li-bog[47] - refers to stingless bees on Panay Island. propolis as "kabulay" in Panay but in Aklan its kabuyay because Aklanon pronounces the letter L as Y. Here in Iloilo, we call *Apis dorsata* as /put-yu-kan/ and apis cerana as /lig-wan/, /li-bog/ for all types of stingless bees.
12. Lukot - Marinduque dialect, some there also call it kiwot or kiyot.
13. Lukot –Tagalog term for stingless bees.
14. Lukutan - or Lulukot here in Sison, Pangasinan Luzon.
15. Mahsam - possibly due to sour SB honey from Iraya Mangyan of Paluan.
16. Mamulit - among Tagbanua of Buenavista in Palawan.
17. Monggo Monggo – Clustered brood SB in Luzon.
18. Mugo Mugo – refers to SB in Guimaras.
19. Pukyutan - stingless bee is called pukyutan in Kalayaan, Laguna.
20. Quiot - in Isabel, Leyte, Visayas it's called "quiot".
21. Takpal - layered brood in Negros Occidental.
22. Tamusan or Laholoy - among the matigsalo Manobo tribe of Bukidnon.
23. Tegteg - stingless bees for Ilocanos in Ilocos Province.
24. Tutoldan - the name of stingless bees in the Cuyonon language of Cuyo Islands, Palawan.

[46] Through Visayas and Davao I found that the Ilonggo or Hiligaynon speaking beekeepers sometimes refer to themselves as 'beeks', although I am not certain if it is a colloquial term or just a slang.
[47] Libog in Filipino actually means "lust" but when it comes to SB, In Romblon it is pronounced "Lib-og" whereas in Panay it is pronounced Li-bog.

DSB 1.1.1 Discussion:

1. There are some similarities to the Malay language, e.g: “Mahsam” in the Ulasan among the Iraya Mangyan of Paluan, Occidental Mindoro where the term means sour, implying the sour attributes of SB honey
2. Interestingly there are also similar-sounding names in Malay-speaking neighbouring countries such as “Tutoldan” in Palawan, “Tantadan” in Sabah, Malaysia and “Teldan” in Lombok, Indonesia.
3. In Malaysia and many parts of Indonesia, “Kelulut” is a common term and we see it also being used as “Kalolot” - Rinconada Bikol
4. Mugo Mugo may be individualistic, a term/name only created by people or groups of people with different pronunciations in their speech.
5. The people of the island province of Guimaras island province speak Hiligaynon (colloquially called Ilonggo) same as in Negros Occidental province on Negros Island and Iloilo province on Panay island. Thus, suggesting a uniform terminology for stingless bees in all Hiligaynon-speaking areas and that is ‘Li-bog’.
6. Monggo-Monggo, on the other hand, applies to the brood cell structure and not to species of stingless bees in general, hence it should not be used to refer to stingless bees. This terminology is new or rather just recently attached to species of stingless bees with clustered brood structures or rather on the brood only. It resembles mongo beans. It therefore may be used nationwide regardless of province and language. Monggo is a poor man's meat and is known by all Filipinos and can easily be used for comparison. If only grapes were much more affordable and common they might have been used for clustered brood, hence the word ubas-ubas. Take note that in Austronesian word structure, word repetition means duplication or pointing to something that resembles the real item. In the Philippines, it is the latter. In addition, monggo-monggo is not a proper native terminology that can be used to address this tiny, stingless tropical eusocial bee (Fernando C. Belmonte Jr., Rizal, Luzon)
7. Panay Island Iloilo City and nearby towns speak Hiligaynon (Ilonggo dialect; the word Ilonggo refers to people from Iloilo speaking Hiligaynon) while the rest of the Province on the island speaks Kinaray-a and can also speak Hiligaynon fluently (Nora Alanza Peña. Iloilo, Panay).
8. When we repeat the word twice, as in monggo monggo, it means tiny or little monggo, balai (house), balai-balai means small house; gamay (little/small) gamay -gamay means very small or very little (Freddie Lozada, Victorias, Negros Occidental)
9. I only want to say that many Filipino languages are based on the Malay language which is why we find strong similarities between the Malaysian and Indonesian languages. Here in Occ.

Mindoro, I find their Tagalog dialect has differences from my Tagalog in central Luzon (region 3) which is the basis of the national language which is Filipino. Theirs are mixed up with other languages and dialects notably the Ilocano and Hiligaynon languages and the southern Tagalog (region 4) dialect (Danny Viola Hizon, Sablayan, Occ. Mindoro).

10. After interviewing locals in the Kalayaan, Laguna they called stingless bees pukyutan almost the same as lukutan. They insisted to me that SBs are called pukyutan in their area although Pukyutan is Tagalog for the giant honeybee. In other words, the SB names may evolve from the vernacular names of honeybees (Bernard Tuzon Anciado, Alaminos, Laguna)
11. Pukyutan is a general term for stingless bees and walang-durong bubuyog is a formal term for stingless bees. In Iloilo, we call *Apis dorsata* put-yu-kan and *Apis cerana* lig-wan/, li-bog for all types of stingless bees (NAP).
12. Kiyot is generally used in Mindanao to refer to stingless bees. However, for the matigsalo Manobo tribe of Bukidnon, the term used for stingless bees is Laholoy or Tamusan. (Datu Makadingding, Bukidnon, Mindanao)
13. In Cebu it is Kiyot (pronounced cute) there is even a barangay in Cebu City named after SB. Quiot, Pardo (Alberto V. Bartolata, Pangdan, Cebu).
14. For Romblon Province 1. Romblon/Sibuyan Island- kabuyay 2. Tablas Island- kiwot, Kiyut In Tablas, kabulay is called propolis in English, for layered brood-Sibuyan- lap-og Tablas (Manakaran dialect)- lib-og. Kabuyay is called by Sibuyanon and Romblomanon. Romblon Province is dominated by three major dialects, 1. Asi dialect 2. Romblomanon dialect 3. The inunhan dialect is locally called the manakaran dialect (Joe Asis, Santa Fe, Tablas, Romblon)
15. Although the Cuyonon language is so closely related to Kinaray-a in Panay, very few Cuyonons live or speak Cuyonon in Panay, they instead settled west to the island of Palawan where the ethnic group is so closely associated now, this being the Province of Palawan declared Cuyonon as its official language. The fact also remains that most of the other ethnic groups of Palawan can fluently speak this language because Cuyonon had been the lingua franca of the Province of Palawan for many centuries already.

    Cuyonon refers to an ethnic group populating the Cuyo Islands, along with northern and central Palawan. The Cuyonons hail originally from Cuyo and the surrounding Cuyo Islands, a group of islands and islets in the northern Sulu Sea, to the northeast of Palawan.

    Ref: https://en.wikipedia.org/wiki/Cuyonon_language
16. In Isabel, Leyte, Visayas it's called "quiot". the pre-Spanish colonization town name was "Quiot" after which it was named Isabel. (Melvin Guerrero, Isabel, Leyte)

Cited Websites

https://en.wikipedia.org/wiki/List_of_Philippine_provincial_name_etymologies

https://www.ilogmaria.com/

https://tagalog.pinoydictionary.com/

https://www.google.com/search?q=binuatan+creations+puerto+princesa+palawan&rlz=1C1RLNS_enPH790PH790&oq=Binuatan+Creations&aqs=chrome.1.69i57j0l7.15681j0j15&sourceid=chrome&ie=UTF-8

https://www.facebook.com/Marinduque-Stingless-Beekeepers-Association-736114896585871/

http://www.discoverlife.org/mp/20q?search=Wallacetrigona&guide=1&flags=HAS:subgenus

http://www.discoverlife.org/mp/20q?search=Tetragonula&guide=1&flags=HAS:subgenus

https://en.wikipedia.org/wiki/Cuyonon_language

https://tl.wikipedia.org/wiki/Wikang_Cuyonon

http://bobthebeeman.com.au/african-tulip-tree-ebook.asp

http://bobthebeeman.com.au/explaining-the-aussie-inpa-box.htm

http://moure.cria.org.br/catalogue

https://en.wikipedia.org/wiki/Beekeeping

https://www.slideshare.net/KeithMikkelson2/apis-trigona-biroi

DSB 1.0 Labelled regional Map of The Philippines

Map 8 Regional Map with vernacular SB names

List of Charts

List of Maps

List of Tables

List of Images

Index

www.ingramcontent.com/pod-product-compliance
Lightning Source LLC
La Vergne TN
LVHW071159210926
848482LV00014B/693